Gas Chromatography and Mass Spectrometry

A Practical Guide

Fulton G. Kitson

Barbara S. Larsen

Charles N. McEwen

ACADEMIC PRESS

San Diego   New York   Boston   London   Sydney   Tokyo   Toronto

This book is printed on acid-free paper. ∞

Copyright © 1996 by ACADEMIC PRESS, INC.

All Rights Reserved.
No part of this publication may be reproduced or transmitted in any form or by any
means, electronic or mechanical, including photocopy, recording, or any information
storage and retrieval system, without permission in writing from the publisher.

Academic Press, Inc.
A Division of Harcourt Brace & Company
525 B Street, Suite 1900, San Diego, California 92101-4495

United Kingdom Edition published by
Academic Press Limited
24-28 Oval Road, London NW1 7DX

Library of Congress Cataloging-in-Publication Data

Kitson, Fulton G.
    Gas chromatography and mass spectrometry : a practical guide / by
Fulton G. Kitson, Barbara S. Larsen, Charles N. McEwen.
        p.   cm.
    Includes bibliographical references and index.
    ISBN 0-12-483385-3 (alk. paper)
    1. Gas chromatography.   2. Mass spectrometry.   I. Larsen, Barbara
Seliger.   II. McEwen, Charles N., date.   III. Title.
QD79.C45K57   1996
543'.0873--dc20   95-47357
                                                       CIP

PRINTED IN THE UNITED STATES OF AMERICA
96   97   98   99   00   01   EB   9   8   7   6   5   4   3   2   1

This book is the culmination of Fulton G. Kitson's highly successful 30-year career in gas chromatography and mass spectrometry. We express our gratitude to Fulton for the many hours he spent after retirement collecting and preparing the documents that are incorporated into this book. We also thank Shirley Kitson for her gracious hospitality during our frequent visits to her home.

Barbara S. Larsen
Charles N. McEwen

To my wife, Shirley, for her encouragement and patience during the preparation of this book.

Fulton G. Kitson

Contents

Preface

Gas Chromatography and Mass Spectrometry: A Practical Guide is designed to be a valuable resource to the GC/MS user by incorporating much of the practical information necessary for successful GC/MS operation into a single source. With this purpose in mind, the authors have kept the reading material practical and as brief as possible. This guide should be immediately valuable to the novice, as well as to the experienced GC/MS user who may not have the breadth of experience covered in this book.

The book is divided into four parts. Part I, "The Fundamentals of GC/MS," includes practical discussions on GC/MS, interpretation of mass spectra, and quantitative GC/MS. Part II, "GC Conditions, Derivatization, and Mass Spectral Interpretation of Specific Compound Types," contains chapters for a variety of compounds, such as acids, amines, and common contaminants. Also included are GC conditions, methods for derivatization, and discussions of mass spectral interpretation with examples. Part III, "Ions for Determining Unknown Structures," is a correlation of observed masses and neutral losses with suggested structures as an aid to mass spectral interpretation. Part IV, "Appendices," contains procedures for derivatization, tips on GC operation, troubleshooting for GC and MS, and other information which are useful to the GC/MS user. Parts I to III also contain references that either provide additional information on a subject or provide information about subjects not covered in this book.

Maximum benefit from *Gas Chromatography and Mass Spectrometry* will be obtained if the user is aware of the information contained in the book. That is, Part I should be read to gain a practical understanding of GC/MS technology. In Part II, the reader will discover the nature of the material contained in each chapter. GC conditions for separating specific compounds are found under the appropriate chapter headings. The compounds for each GC separation are listed in order of elution, but more important, conditions that are likely to separate similar compound types are shown. Part II also contains information on derivatization, as well as on mass spectral interpretation for derivatized and underivatized compounds. Part III, combined with information from a library search, provides a list of ion masses and neutral losses for interpreting unknown compounds. The appendices in Part IV contain a wealth of information of value to the practice of GC and MS.

The GC separations, derivatization procedures, mass spectral interpretation, structure correlations, and other information presented in this book were collected or experimentally produced over the length of a 30-year career (F.G.K.) in GC/MS. It has not been possible to reference all sources; therefore, in the acknowledgments, we thank those persons whose work has significantly influenced this publication.

Fulton G. Kitson
Barbara S. Larsen
Charles N. McEwen

Acknowledgments

Special thanks to Alfred Bolinski, Willard Buckman, Iwan deWit, James Farley, Richard McKay, Raymond Richardson, and Francis Schock for the efforts, skills, and good humor they brought to the GC/MS laboratory. William Askew and Albert Ebert were managers in the early days (1960s) who were key to advancing the science and who suggested that we maintain notes on separations and mass/structure correlations, which are the foundations of this book. Over the years, fruitful discussions with Al Beattie, Bernard Lasoski, Dwight Miller, Daniel Norwood, Thomas Pugh, Robert Reiser, James Robertson, Pete Talley, Rosalie Zubyk, and others have provided concepts and methods that are incorporated into this book.

We are indebted to the authors whose works have influenced this guide but may not be appropriately referenced. They are B. A. Anderson, S. Abrahmsson, J. H. Beynon, K. Bieman, C. J. Bierman, J. C. Cook, R. G. Cooks, D. C. Dejongh, C. Djerassi, C. C. Fenselau, J. C. Frolich, R. L. Foltz, N. G. Foster, B. J. Gudzinowicz, W. F. Haddon, A. Harrison, M. C. Hamming, D. R. Knapp, J. A. McCloskey, S. MacKenzie, F. W. McLafferty, D. S. Millington, H. F. Morris, B. Munson, S. Meyerson, K. Pfleger, R. I. Reed, V. N. Reinhold, F. Rowland, R. Ryhage, B. E. Samuelson, J. Sharkey, S. Shrader, E. Stenhagen, R. Venkataraghaven, and J. T. Watson.

We thank the DuPont Company for providing the resources for the preparation of this book. VG Fisons and NIST also have graciously permitted the use of spectra from their mass spectral libraries. Roger Patterson, as well, is gratefully acknowledged for assisting in the electronic transfer of the spectra presented in this guide.

The authors also are indebted to Mrs. Phyllis Reid for her skills in desktop publishing, ChemDraw, and other computer software programs, but mostly for her perseverance and good humor during the many revisions of this book. Phyllis' input has improved the quality of this product and her knowledge has allowed us to put the entire manuscript into an electronic format.

Part I

The Fundamentals of GC/MS

C h a p t e r   1

What Is GC/MS?

Gas chromatography/mass spectrometry (GC/MS) is the synergistic combination of two powerful analytic techniques. The gas chromatograph separates the components of a mixture in time, and the mass spectrometer provides information that aids in the structural identification of each component. The gas chromatograph, the mass spectrometer, and the interface linking these two instruments are described in this chapter.

The Gas Chromatograph

The gas chromatograph was introduced by James and Martin in 1952.[1] This instrument provides a time separation of components in a mixture. The basic operating principle of a gas chromatograph involves volatilization of the sample in a heated inlet port (injector), separation of the components of the mixture in a specially prepared column, and detection of each component by a detector. An important facet of the gas chromatograph is the use of a carrier gas, such as hydrogen or helium, to transfer the sample from the injector, through the column, and into the detector. The column, or column packing, contains a coating of a stationary phase. Separation of components is determined by the distribution of each component between the carrier gas (mobile phase) and the stationary phase. A component that spends little time in the stationary phase will elute quickly. Only those materials that can be vaporized without decomposition are suitable for GC analysis. Therefore, the key features of gas chromatographs are the systems that heat the injector, detector, and transfer lines, and allow programmed temperature control of the column.

Carrier Gas

Helium is generally the carrier gas, but hydrogen and nitrogen are often used in certain applications. The carrier gas must be inert and cannot be adsorbed by the column stationary phase. An important parameter is the linear velocity of the carrier gas. For helium, 30 cm/sec is optimum. The linear velocity can be determined by injecting a compound, such as argon or butane, that is not retarded by the column stationary phase and measuring the time from injection to detection. Hence, the linear velocity is the retention time in seconds divided into the column length in centimeters. The pneumatics must be capable of providing a stable and reproducible flow of carrier gas.

Sample Introduction

There are several types of sample introduction systems available for GC analysis. These include gas sampling valves, split and splitless injectors, on-column injection systems, programmed-temperature injectors, and concentrating devices. The sample introduction device used depends on the application.

Gas Sampling Valves: The gas sampling valve is used for both qualitative and quantitative analyses of gases. The valve contains a loop of known volume into which gas can flow when the valve is in the sampling position. By changing the valve to the analyzing position, the gas in the loop is transferred by the carrier gas into the GC column. Gas sampling valves can be operated at reduced pressure for analysis of low-boiling liquids that vaporize at reduced pressures.

Split Injection[2]:   In the split injector, the injected sample is vaporized into the stream of carrier gas, and a portion of the sample and solvent, if any, is directed onto the head of the GC column. The remainder of the sample is vented. Typical split ratios range from 10:1 to 100:1 and can be calculated from the equation:

$$\text{Split ratio} = \frac{\text{column flow} + \text{vent flow}}{\text{column flow}}$$

where the approximate

$$\text{Column flow} = \frac{\pi \, (\text{internal column radius in cm})^2 (\text{column length in cm})}{(\text{retention time of argon or butane in min})}$$

Column flow example:

$$\text{Column flow} = \frac{(3.141) \times (0.025 \text{ cm})^2 \times 3000 \text{ cm}}{2.50 \text{ min}} = 2.36 \text{ cm}^3/\text{min}$$

Normally, 1–2 μl of sample is injected into a split-type injector, but larger volumes (3–5 μl) can also be used.

Splitless Injection[2]: In splitless injection, the splitter vent is closed so that all of the sample flows onto the head of the column. After a specific time called the purge activation time, the splitter vent is opened to purge solvent from the injector and any low-boiling components of the sample that are not adsorbed by the column. Splitless injection, therefore, concentrates the sample onto the head of the cool column and purges most of the volatile solvent. For this reason, and because large amounts of sample can be injected, splitless injection is used for trace analysis. The splitless method is not recommended for wide-boiling range samples if quantitation is required. For best results, the solvent boiling point should be at least 20° below the lowest boiling component of the sample. Although splitless injection is the preferred method for trace analyses, it does require optimization of such parameters as column temperature and purge time.

On-column Injection: With on-column injection, the sample is injected directly onto the column using a small syringe needle. Obviously, this technique is easier to use with larger bore GC columns, but modern gas chromatographs can precisely control the on-column injection process, including automatic control of heating and cooling of the injector. This method of analysis gives good quantitative results and is especially valuable for wide-boiling ranges and thermally labile samples. With this technique, a short section of uncoated (inert) fused silica capillary tubing is often inserted between the injection port and the capillary analytic column.

Programmed Temperature Injectors: The programmed temperature injector is held near the boiling point of the solvent; after injection of the sample, it is temperature programmed rapidly until it reaches the desired maximum temperature, which is normally higher than that of an isothermal (constant temperature) injector. As the sample components vaporize, they are transferred onto the head of the GC column. This technique is a variation of on-column injection, but reduces the peak broadening frequently seen with on-column injections.

Concentrating Devices for Sample Injection: Several concentrating devices for organic chemical analyses are commercially available. These devices interface with the inlet system of the gas chromatograph and concentrate organics from large samples of air or water. Most of these devices trap the organics onto adsorbents such as charcoal and/or porous polymers. The sample is thermally desorbed onto the head of a GC column by reverse flushing with the carrier gas. Concentration devices are also used for analyzing off-gases from such materials as polymers. Often, a simple cool stage is sufficient to trap volatiles that are subsequently desorbed by rapidly increasing the temperature of the trapping device.

GC Columns

In a gas chromatograph, separation occurs within a heated hollow tube, the column. The column contains a thin layer of a nonvolatile chemical that is either coated onto the walls of the column (capillary columns) or coated onto an inert solid that is then added to the column (packed columns). The components of the injected sample are carried onto the column by the carrier gas and selectively retarded by the stationary phase. The temperature of the oven in which the GC column resides is usually increased at 4°–20°/minute so that higher boiling and more strongly retained components are successively released. Gas chromatography is limited to compounds that are volatile or can be made volatile and are sufficiently stable to flow through the GC column. Derivatization can be used to increase the volatility and stability of some samples. Acids, amino acids, amines, amides, drugs, saccharides, and steroids are among the compound classes that frequently require derivatization. (See Appendix 3 for procedures used to derivatize compounds for analysis by GC/MS.)

Stationary Phases: The best general purpose phases are dimethylsiloxanes (DB-1 or equivalent) and 5% phenyl/95% dimethylsiloxane (DB-5 or equivalent). These rather nonpolar phases are less prone to bleed than the more polar phases. The thickness of the stationary phase is an important variable to consider. In general, a thin stationary phase (0.3 μm) is best for high boilers and a thick stationary phase (1.0 μm) provides better retention for low boilers. (For more detailed information, see "Stationary Phase Selection" in Appendix 2.)

GC Detectors

One great advantage of GC is the variety of detectors that are available. These include universal detectors, such as flame ionization detectors and selective detectors, such as flame photometric and thermionic detectors. The most generally useful detectors, excluding the mass spectrometer are described in the following sections.

Flame Ionization Detector: The analyte in the effluent enters the flame ionization detector (FID) and passes through a hydrogen/air flame. Ions and electrons formed in the flame cause a current to flow in the gap between two electrodes in the detector by decreasing the gap resistance. By amplifying this current flow a signal is produced. Flame ionization detectors have a wide range of linearity and are considered to be universal detectors even though there is little or no response to compounds such as oxygen, nitrogen, carbon disulfide, carbonyl sulfide, formic acid, hydrogen sulfide, sulfur dioxide, nitric oxide, nitrous oxide, nitrogen dioxide, ammonia, carbon monoxide, carbon dioxide, water, silicon tetrafluoride, silicon tetrachloride, and others.

Thermal Conductivity Detector: In the thermal conductivity detector (TCD), the temperature of a hot filament changes when the analyte dilutes the carrier gas. With a constant flow of helium carrier gas, the filament temperature will remain constant, but as compounds with different thermal conductivities elute, the different gas compositions cause heat to be conducted away from the filament at different rates, which in turn causes a change in the filament temperature and electrical resistance. The TCD is truly a universal detector and can detect water, air, hydrogen, carbon monoxide, nitrogen, sulfur dioxide, and many other compounds. For most organic molecules, the sensitivity of the TCD detector is low compared to that of the FID, but for the compounds for which the FID produces little or no signal, the TCD detector is a good alternative.

Thermionic Specific Detector: The thermionic specific detector (TSD) is similar to the FID with the addition of a small alkali salt bead, such as rubidium, which is placed on the burner jet. Nitrogen and phosphorus compounds increase the current in the plasma of vaporized metal ions. The detector can be optimized for either nitrogen-containing compounds or phosphorus-containing compounds by carefully controlling of the bead temperature and hydrogen and air flow rates. The detector can be tuned using azobenzene for nitrogen and parathion for phosphorus.

Flame Photometric Detector[3]**:**   With the flame photometric detector (FPD), as with the FID, the sample effluent is burned in a hydrogen/air flame. By using optical filters to select wavelengths specific to sulfur and phosphorus and a photomultiplier tube, sulfur or phosphorus compounds can be selectively detected.

Electron Capture Detector: In the electron capture detector (ECD), a beta emitter such as tritium or ^{63}Ni is used to ionize the carrier gas. Electrons from the ionization migrate to the anode and produce a steady current. If the GC effluent contains a compound that can capture electrons, the current is reduced because the resulting negative ions move more slowly than electrons. Thus, the signal measured is the loss of electrical current. The ECD is very sensitive to materials that readily capture electrons. These materials frequently have unsaturation and electronegative substituents. Because the ECD is sensitive to water, the carrier gas must be dry.

The GC/MS Interface

The interface in GC/MS is a device for transporting the effluent from the gas chromatograph to the mass spectrometer. This must be done in such a manner that the analyte neither condenses in the interface nor decomposes before entering the mass spectrometer ion source. In addition, the gas load entering the ion source must be within the pumping capacity of the mass spectrometer.

Capillary Columns

For capillary columns, the usual practice is to insert the exit end of the column into the ion source. This is possible because under normal operating conditions the mass spectrometer pumping system can handle the entire effluent from the column. It is then only necessary to heat the capillary column between the GC and the MS ion source, taking care to eliminate cold spots where analyte could condense. The interface must be heated above the boiling point of the highest-boiling component of the sample.

Macrobore and Packed Columns

The interface for macrobore and packed columns is somewhat more complicated than that for capillary columns because the effluent from these columns must be reduced before entering the ion source. Splitting the effluent is not satisfactory because of the resulting loss of sensitivity. Instead, enrich-

ment devices are used. The most common enrichment device used in GC/
MS is the jet separator.

Jet Separator: The jet separator contains two capillary tubes that are
aligned with a small space (*ca.* 1 mm) between them. A vacuum is created
between the tubes by using a rotary pump. The GC effluent passes through
one capillary tube into the vacuum region. Those molecules that continue
in the same direction will enter the second capillary tube and will be directed
to the ion source. Enrichment occurs because the less massive carrier gas
(He) atoms are more easily collisionally diverted from the linear path than
the more massive analyte molecules.

As with capillary columns, it is crucial to have an inactive surface and
maintain a reasonably even temperature over the length of the interface.
This is usually accomplished by using only glass in the interface. The additional connections necessary in an enrichment-type interface present new
areas for leaks to occur. Connections are especially prone to develop leaks
after a cooling/heating cycle.

The Mass Spectrometer

In 1913, J. J. Thomson[4] demonstrated that neon consists of different atomic
species (isotopes) having atomic weights of 20 and 22 g/mole. Thomson is
considered to be the "father of mass spectrometry." His work rests on
Goldstein's (1886) discovery of positively charged entities and Wein's
(1898) demonstration that positively charged ions can be deflected by electrical and magnetic fields.

A mass spectrometer is an instrument that measures the mass-to-charge
ratio (m/z) of gas phase ions and provides a measure of the abundance of
each ionic species. The measurement is calibrated against ions of known
m/z. In GC/MS, the charge is almost always 1, so that the calibrated scale
is in Daltons or atomic mass units. All mass spectrometers operate by
separating gas phase ions in a low pressure environment by the interaction
of magnetic or electrical fields on the charged particles. The most common
mass spectrometers interfaced to gas chromatographs are the so-called
quadrupole and magnetic-sector instruments.[5]

Magnetic-Sector Instruments

In the magnetic-sector instrument (Figure 1.1), gas phase ions produced in
the ion source by one of several different methods are accelerated from
near rest (thermal energy) through a potential gradient (commonly kV).
These ions travel through a vacuum chamber into a magnetic field at a

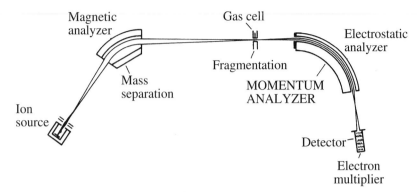

Figure 1.1.   Schematic of a double-focusing reverse geometry magnetic-sector instrument.

sufficiently low pressure such that collisions with neutral gas molecules are uncommon. All of the ions entering the magnetic field have approximately the same kinetic energy (eV). A magnetic field, B, exerts a force perpendicular to the movement of the charged particles according to the equation $m/z = B^2 r^2/2V$, where r is the radius of curvature of the ions traveling through the magnetic field. Thus, at a given magnetic field and accelerating voltage, ions of low mass will travel in a trajectory having a smaller radius than the higher mass ions. In practice, the ions have to pass through a fixed slit before striking a detector. Sweeping the magnetic field from high to low field causes ions of successively lower mass to pass through the slit and strike the detector. This pattern of detected ion signals, when displayed on a calibrated mass scale, is called a mass spectrum.

Electrostatic Analyzer: In magnetic-sector instruments, an electrostatic sector can be incorporated either before or after the magnet to provide energy resolution and directional focusing of the ion beam. The resolution achievable in these double-focusing instruments is sufficient to separate ions having the same nominal mass (e.g., 28 Daltons) but with different chemical formula (e.g., N_2 and CO).

The electrostatic sector is constructed of two flat curved metal plates having opposite electrical potentials. Positive ions traveling between the plates are repelled by the plate with positive potential toward the plate of negative potential. The potential on the plates is adjusted so that ions having eV-kinetic (translational) energy, which is determined by the accelerating voltage of the ion source, will follow the curvature of the plates. Slight differences in the translational energies of the ions (due to the Boltzman distribution and field inhomogeneities in the ion source) are compensated by the velocity-focusing properties of the electrostatic field. The electrostatic sector improves both the mass resolution and stability of the mass spectrometer.

Resolution and Mass Accuracy: With a modern double-focusing mass spectrometer, it is possible to measure the mass of an ion to one part per million (ppm) or better and obtain 100,000 or better resolution. Under GC/MS scanning conditions, 5 to 10 ppm mass accuracy is more common and resolution is set between 2000 and 10,000 ($m/\Delta m$, 10% valley). This mass accuracy is often sufficient in GC/MS analyses to allow for only a few reasonable and possible elemental compositions. For example, if the mass of an ion is determined to be 201.115, and it is expected from other information that only carbon, hydrogen, oxygen, and nitrogen atoms are present in the ion, then within a 15 ppm mass window, there are only three reasonable elemental compositions. If this ion is known to be a molecular ion, only the elemental composition $C_{13}H_{15}NO$ has a whole number of rings and double bonds and follows the Nitrogen Rule (see Chapter 2). A knowledge of the elemental composition of the molecular ion and fragment ions greatly simplifies interpretation.

Resolution is necessary in accurate mass measurement to eliminate ions from mass analysis that have the same nominal mass (e.g., 201) but different elemental compositions and thus a small mass difference (e.g., 201.115 and 201.087). The resolution necessary to separate these ions is calculated from the formula $Res = m/\Delta m = 201.087/(201.115 - 201.087) = 7500$. Resolution is defined as the separation of the ion envelope of two peaks of equal intensity differing in mass by Δm. If the ion envelopes (tops of the peaks) are separated by approximately 1.4 times the width of the peaks at half height, then the ion envelopes will overlap with a 50% valley (see Figure 1.2). The resolution is then defined as $m/\Delta m$, 50% valley, where m is the mass of the lower mass peak that is being resolved and the 50% point is measured from the baseline to the crossover point of the peaks. With magnetic-sector instruments, a 5 or 10% valley is frequently specified. Obviously, an instrument will need increasingly higher resolving power to achieve a resolution of 7500 with a 50, 10, or 5% valley, respectively.

In a magnetic-sector instrument, resolution is increased by restricting the height and width of the ion beam and by tuning using electrical lenses. The most important resolution effect is obtained from adjustable slits just outside the source region and just before the detector, which restrict the width (y dispersion) of the ion beam. Increasing the resolution attenuates the beam; hence, for accurate mass analyses using GC/MS, a compromise must be made between the resolution necessary to minimize mass interference and the signal intensity necessary to detect low levels of material.

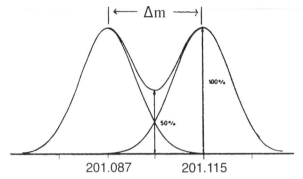

Figure 1.2.   Resolution as a function of peak width.

Because the gas chromatograph eliminates most compounds that will cause mass interference, the principle cause of peak overlap is the reference material used as an internal mass standard. The most common reference material for accurate mass measurement is perfluorokerosine (pfk), which because of the number of fluorine atoms in each molecule has a negative mass defect. Because there are numerous (reasonably) evenly spaced fragment and molecular ions, this material is a good reference for obtaining high mass accuracy. Unfortunately, pfk requires high instrument resolution (5000–10,000) to eliminate interferences with ions from the compounds whose mass will be measured. Alternatively, compounds with large mass defects can be used as internal mass references. For example, perfluoroiodo-compounds have such large negative mass defects that 2000 resolution (10% valley) is sufficient to eliminate almost all compound/reference interferences. The lower mass resolution results in greater sensitivity.

Quadrupole Instruments

The other common instrument for GC/MS analysis is the quadrupole mass filter. These instruments derive their name from the four precisely machined rods that the ions must pass between to reach the detector. Ions enter the quadrupole rods along the z-axis after being drawn out of the ion source by a potential (typically a few volts). The ions entering the quadrupole are sorted by imposing rf and dc fields on diagonally opposed rods. By sweeping the rf and dc voltages in a fixed ratio, usually from low to high voltages, ions of successively higher masses follow a stable path to the detector. At any given field strength, only ions in a narrow m/z range reach the detector. All others are deflected into the rods.

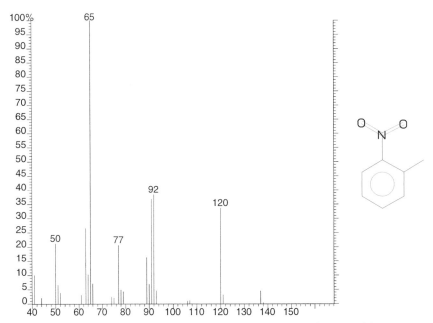

Figure 1.3. Mass spectrum showing a small molecular ion at *m/z* 137 and lower mass fragment ions.

Resolution: Quadrupole instruments are not capable of achieving the high resolution that is common with double-focusing magnetic-sector instruments. In GC/MS analyses, a compromise is struck between sensitivity (ion transmission) and mass resolution. In the quadrupole instrument, the resolution is set to the lowest possible value commensurate with resolving peaks differing by 1 Dalton (unit resolution).

The Mass Spectrum

A mass spectrum is a graphic representation of the ions observed by the mass spectrometer over a specified range of *m/z* values. The output is in the form of an *x,y* plot in which the *x*-axis is the mass-to-charge scale and the *y*-axis is the intensity scale. If an ion is observed at an *m/z* value, a line is drawn representing the response of the detector to that ionic species. The mass spectrum will contain peaks that represent fragment ions as well as the molecular ion (see Figure 1.3). Interpretation of a mass spectrum identifies, confirms, or determines the quantity of a specific compound.

Both the intensity and *m/z* axis are important in interpreting a mass spectrum. The relative intensity of ions that are observed in a mass spectrum of a pure compound may vary from instrument to instrument. In general, most magnetic-sector instruments will produce very similar spectra, provided the ionization conditions are the same. Quadrupole instruments are tuned to produce mass spectra that are similar to spectra obtained from magnetic-sector instruments. In either case, properly calibrated instruments will have all the same ions at the same m/z values. Thus, two mass spectra taken over the same mass range using the same ionization conditions should show all the same ions with some variation in the relative intensity of observed ions. Extra peaks appearing in a spectrum are caused by impurities or background peaks.

Because the vacuum in the mass spectrometer and the cleanliness of the ion source, transfer line, GC column, and so forth are not perfect, a mass spectrum will typically have several peaks that are due to background. All GC/MS spectra, if scanned to low enough mass values, will have peaks associated with air, water, and the carrier gas. Other ions that are observed in GC/MS are associated with column bleed and column contamination.

Multisector Mass Spectrometers[6]

Mass analyzers can be combined in a tandem arrangement to give additional information. Tandem mass spectrometers are referred to as MS/MS instruments. The most common of these are the triple quadrupole instruments that use two sets of quadrupole rods for mass analysis, connected in a tandem arrangement by a third rf-only set of quadrupole rods, which act to transmit ions that undergo collisional fragmentation. Magnetic-sector instruments are also combined in tandem. Three- and four-sector instruments are commercially available. These can be combined as BEB, BEBE, or BEEB, where B refers to the magnetic sector and E refers to the electrostatic sector. Hybrid BEQ (quadrupole) and BE/time-of-flight (TOF) mass spectrometers are also available.

Instruments are available that can perform MS/MS type experiments using a single analyzer. These instruments trap and manipulate ions in a trapping cell, which also serves as the mass analyzer. The ion trap and fourier transform ion cyclotron resonance (FT-ICR) mass spectrometers are examples.

MS/MS Instrumentation: As was noted previously, a variety of instrument types can perform MS/MS experiments, but because of their popularity, we only discuss MS/MS experiments using triple quadrupole instruments. The principles can be applied to other types of instrumentation.

The triple quadrupole instrument consists of two mass analyzers separated by an rf-only quadrupole. In the rf-only mode, ions of all masses are transmitted through the quadrupole filter. By adding a gas such as argon into the space between the rf-only rods, ions entering this space from the first mass filter can undergo multiple collisions with the neutral argon atoms. If the ion-neutral collisions are sufficiently energetic, fragmentation of the ions will result. These fragment ions pass through the rf-only quadrupole, and their masses are measured using the final quadrupole mass filter.

The rf and dc voltages on the first quadrupole mass filter can be set to allow ions of a selected mass to enter the rf-only quadrupole and undergo collisional fragmentation. The masses of the fragment ions are determined by scanning the third quadrupole. In this way, the fragment (product/daughter) ions of a selected precursor (parent) ion can be determined. Alternatively, by scanning the first set of quadrupole rods and setting the third set to pass a given fragment (product) ion, all of the precursor ions of the selected fragment ions can be identified. (Fragment ions will be observed only when their precursor ions are transmitted by the first set of quadrupole rods.) Scanning the first and third quadrupoles with a fixed mass difference identifies neutral losses. These methods are very powerful for identifying unknowns and for providing additional selectivity when searching for known compounds in complex mixtures.

Other Analyzer Types

There are other types of analyzers used for GC/MS that are not yet as common as quadrupole and magnetic-sector instruments. FT-ICR, TOF, and quadrupole ion trap mass spectrometers are important in certain applications. The FT-ICR instrument can provide very high resolution and MS/MS capabilities. The TOF instrument is advantageous for rapid output or high-resolution GC where rapid acquisition is required. The quadrupole ion trap mass spectrometer was first developed as a GC detector with mass selection capabilities, but considerable improvements in the last several years are increasing the importance of these instruments in GC/MS analyses. High sensitivity and MS/MS capabilities, along with the potential for high resolution, are important advantages of this instrument.

Ion Detection

Detecting ions in GC/MS is performed almost exclusively using an electron multiplier. There are two types of electron multipliers: the continuous dynode type and the discrete type. Both operate on the principle that ions with sufficient kinetic energy will emit secondary electrons when they strike a metal surface. The discrete type of electron multiplier has a series of dynodes (metal plates that look like partially open venetian blinds) that are connected by a resistor chain so that the first dynode has a higher negative potential than the last dynode. The electrons emitted from the first dynode are accelerated through a sufficiently high voltage gradient to cause multiple electron emission when they strike the surface of the second dynode. Repeating this process will result in an increasing cascade of electrons that, at the end of the cascade, will provide sufficient signal to be detected. The amplified signal is then sent to a computer or other output device for processing. The continuous dynode electron multiplier is more frequently used in quadrupole instruments and operates on the same principle as the discrete dynode multiplier, except that it has a continuous curved surface over which there is a voltage drop. The electrons that are emitted when ions strike the front of the dynode will be accelerated toward the rear, but will collide with the curved surface causing multiplication of the signal.

The electron multiplier is well suited for the detection of positive ions because the first dynode can be set at a negative potential and the final dynode can be at ground potential. The positive ions are then accelerated toward the first dynode and the secondary electrons toward ground (more positive). For negative ion detection, the situation is reversed. If the signal dynode is to be at ground potential, the first dynode must be at a high negative potential for electron amplification. In this case, negative ions decelerate before striking the first dynode. This problem can be overcome by placing a conversion dynode before the electron multiplier. The conversion dynode can be biased negative relative to the first dynode for positive ion detection, so that electrons emitted from the surface of the conversion dynode upon ion impact will be attracted to the first multiplier dynode, and normal amplification will occur. To detect negative ions, the conversion dynode is operated at a positive potential, and the first dynode of the multiplier is kept at the normal negative potential. When ions strike the conversion dynode, electrons and ions are emitted from its surface. The positive ions are attracted to the negative first dynode, thus beginning the process of signal amplification.

Ionization Methods

There are numerous ionization techniques available to the mass spectrometrist, but for GC/MS almost all analyses are performed using either electron impact ionization or chemical ionization.

Electron Impact Ionization: Electron impact ionization (ei) is by far the most commonly used ionization method. The effluent from the GC enters a partially enclosed ion source. Electrons "boiled" from a hot wire or ribbon (filament) are accelerated typically by 70 V (and thus have 70 eV of energy) before entering the ion source through a small aperature. When these electrons pass near neutral molecules, they may impart sufficient energy to remove outer shell electrons, producing additional free electrons and positive (molecular) ions. The energy imparted by this type of ionization is high and, with only rare exceptions, causes part of or all of the molecular ions to break apart into neutral atoms and fragment ions. This ionization technique produces almost exclusively positively charged ions.

$$M + e^- \longrightarrow M^{+\cdot} + 2e^- \tag{1}$$

$$M^{+\cdot} \longrightarrow F^+ + N^\cdot \tag{2}$$

Chemical Ionization[7]: Chemical ionization (ci), like ei, generates ions using an electron beam. The primary difference is that the ion chamber used for the ionization is more tightly closed than that used in ei so that a higher pressure of gas can be added to the chamber while maintaining a good vacuum along the ion flight path. Several different gases have been used in ci, but for illustrative purposes, only methane is discussed in detail.

Addition of methane to the ion source at a pressure of about 0.5 Torr causes almost all of the electrons entering the ion source to collide with methane molecules. The first event is the expected production of a molecular ion (eq. 3). The molecular ion can then undergo fragmentation (eq. 4) or because of the high pressure of neutral methane, ion-molecule reactions can occur (eqs. 5 and 6).

$$CH_4 + e^- \longrightarrow CH_4^{+\cdot} + 2e^- \tag{3}$$

$$CH_4^{+\cdot} \longrightarrow CH_3^+, CH_2^{+\cdot}, \text{etc.} \tag{4}$$

$$CH_4^{+\cdot} + CH_4 \longrightarrow CH_5^+ + \cdot CH_3 \tag{5}$$

$$CH_3^+ + CH_4 \longrightarrow C_2H_5^+ + H_2 \tag{6}$$

$$CH_5^+ + M \longrightarrow CH_4 + MH^+ \tag{7}$$

The result of the fast reactions in the ion source is the production of two abundant reagent ions (CH_5^+ and $C_2H_5^+$) that are stable in the methane plasma (do not react further with neutral methane). These so-called reagent ions are strong Brønsted acids and will ionize most compounds by transferring a proton (eq. 7). For exothermic reactions, the proton is transferred from the reagent ion to the neutral sample molecule at the diffusion controlled rate (at every collision, or *ca.* 10^{-9} s^{-1}).

Unlike ei, ci usually produces even-electron protonated $[M + H]^+$ molecular ions. Usually, less fragmentation and more abundant molecular ions are produced with ci because less energy is transferred during ionization and because even-electron ions are inherently more stable than the odd-electron counterparts produced by ei. By producing weaker Brønsted acids as reagent ions, less energy will be transferred during ionization of the analyte (see Table 1.1). This can be done, for example, by substituting isobutane for methane as the reagent gas to produce the weaker Brønsted acid, $t\text{-}C_4H_9^+$. Alternatively, a small amount of ammonia can be added to methane to produce the NH_4^+ reagent ion. NH_4^+ will only transfer a proton to compounds that are more basic (have a higher gas phase proton affinity) than ammonia. In ci, the reagent ion can also form adduct ions with sample molecules (e.g., $[M + NH_4^+]$).

Positive ion ci is useful in GC/MS when more intense molecular ions and less fragmentation are desirable. However, in some molecules, such as saturated alcohols, protonation occurs at the functional group, which in the case of alcohols, results in efficient loss of H_2O from the molecule; thus, if a molecular ion is observed, it is very small. Another potential advantage of ci is the ability to tailor the reagent gas to the problem. For example, if one is only interested in determining which amines are present in a complex hydrocarbon mixture, ammonia would be a good choice for a reagent gas because the NH_4^+ reagent ion is not acidic enough to protonate hydrocarbon molecules, but will protonate amines.

Negative Ion Chemical Ionization: Negative ions are produced under ci conditions by electron capture. Under the higher pressure conditions of the ci ion source, electrons, both primary (those produced by the filament) and secondary (produced during an ionization event), undergo collisions until they reach near-thermal energies. Under these conditions, molecules

Table 1.1.   Proton affinities for selected reagent gases

Reagent gas	Reagent ions	Proton affinity*
H_2	H_3^+	101
CH_4	CH_5^+	132
H_2O	H_3O^+	167
$i\text{-}C_4H_{10}$	$i\text{-}C_4H_9^+$	196
NH_3	NH_4^+	204

*The higher the proton affinity, the weaker the Brønsted acid.

with high electron affinities (frequently containing electronegative substituents) are able to capture electrons very efficiently. For certain types of molecules, this is a very sensitive ionization method because the electron-molecule collision rate is much faster than the rate for ion-molecule collisions. This faster collision rate is due to the higher rate of diffusion of electrons versus the more massive ions. Negative ions can also be produced from negative reagent ions, but this method is inherently less sensitive than electron capture and infrequently used in GC/MS.

Negative ion ci is often used to analyze highly halogenated, especially fluoronated molecules as well as other compounds containing electronegative substituents. Completely saturated compounds such as perfluoroalkanes have only antibonding orbitals available for electron capture, and these molecules undergo dissociative electron capture, often producing abundant fragment ions. Molecules containing both electron-withdrawing substituents and unsaturation, such as hexafluorobenzene, readily capture electrons to produce intense molecular ions. Because electron capture is so efficient for this kind of molecule, and because of the high rate of electron-molecule collisions, negative ion electron capture MS can be as much as 10^3 times more sensitive than positive ion ci.

Selected Ion Monitoring

Selected ion monitoring (SIM) refers to the use of the instrument to record the ion current at selected masses that are characteristic of the compound of interest in an expected retention time window. In this mode, the mass spectrometer does not spend time scanning the entire mass range, but rapidly changes between m/z values for which characteristic ions are expected. The SIM method allows quantitative analysis at the parts per billion (ppb) level. With modern instruments, the data system can be programmed to examine different ions in multiple retention time windows. The advantage of this method is that both high sensitivity and high specificity are achieved.

A typical example of the use of SIM is the quantitative determination of certain specific compounds in a complex mixture, especially when the compounds are present at low levels. Although SIM is sensitive to picograms of material, this sensitivity is highly dependent on the matrix containing the compounds of interest and the interferences that are produced. Frequently, it is only possible to detect nanogram levels of the compound because of the chemical interferences. It is not unusual to obtain erratic results when using small amounts of material because interfering ions are within the mass window of the selected ion monitoring experiment even when internal standards are employed. If the appropriate instrumentation is available, it is often possible to reduce or eliminate the background interference by increasing the resolution and thus narrowing the mass window. This method results in some loss of signal intensity associated with obtaining increased mass resolution; however, it is often accompanied by an increase in the signal-to-noise ratio due to less background interference. An alternative way to eliminate background interference is to use MS/MS combined with SIM to monitor characteristic fragment ions.

MS/MS and Collision-Induced Dissociation

Tandem quadrupole and magnetic-sector mass spectrometers as well as FT-ICR and ion trap instruments have been employed in MS/MS experiments involving precursor/product/neutral relationships. Fragmentation can be the result of a metastable decomposition or collision-induced dissociation (CID). The purpose of this type of instrumentation is to identify, qualitatively or quantitatively, specific compounds contained in complex mixtures. This method provides high sensitivity and high specificity. The instrumentation commonly applied in GC/MS is discussed under the "MS/MS Instrumentation" heading, which appears earlier in this chapter.

MS/MS can be used for the same type of experiments described previously for SIM, but provides increased selectivity and essentially eliminates chemical background interference. A common experiment is to determine the approximate amounts of specific compounds in a very complex matrix for which either the best GC conditions could not resolve all the peaks or the compound of interest is obscured by chemical noise (background). In this experiment, the first mass filter can be set to pass selected precursor ions (during a time window) at approximately the expected retention time for each component. When the compound elutes from the GC, the precursor ions will pass through the first mass filter while ions of all other masses will not. These precursor ions then enter a special collision region, into which a gas is introduced, and collide with the gas molecules. If the ions are given sufficient energy during the collisional process, they will produce characteristic fragment ions that enter the second mass filter. Depending on the nature of the experiment, the second mass analyzer can be scanned to observe all product ions, can jump from peak to peak to spend more time on specific product ions, or can be set to a constant field to select a single specified product ion. Although the specificity decreases from the scanned mode to the constant field mode, the sensitivity increases.

Isotope Peaks

Isotope peaks can be very informative in GC/MS analysis. Generally for interpretation, one focuses on the monoisotopic peak. The monoisotopic

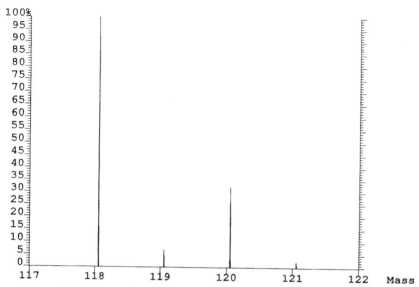

Figure 1.4. Isotope cluster for $C_6H_{11}Cl$. The monoisotopic peak is at *m/z* 118.

peak is the lowest mass peak in an isotope cluster and is comprised of the lowest mass isotope of each element in the ion (see Figure 1.4). However, because most elements, including carbon, have isotope peaks, the ions observed in mass spectra will have isotopes that are characteristic of their elements. Therefore, good isotope ratios can provide information concerning the elemental composition of the compound being analyzed. The elements of chlorine, bromine, sulfur, and others will stand out when examining a mass spectrum because of the high intensity of their isotope peaks. Because carbon has a ^{13}C isotope that is 1.1% the abundance of the ^{12}C peak, it is sometimes possible to determine the number of carbon atoms in a molecule. For example, a molecule having 10 carbon atoms will have a peak that is 1 Dalton higher in mass than the monoisotopic peak and has an abundance that is 11% (1.1×10) of the monoisotopic peak. (For more information on using isotope patterns to determine elemental composition, see Chapter 2.)

Metastable Ions[8]

Metastable ions are almost always the result of fragmentation after a rearrangement process. The reason for this is that direct fragmentation is prompt, occurring before the ions leave the ion source. Metastable ions must survive the ion source and the region of ion acceleration just outside the ion source before fragmenting in a field-free region of the mass spectrometer. For ions that fragment after reaching full ion acceleration, the kinetic energy of the fragment ions is represented by $E = \dfrac{M_f eV}{M_p}$, where M represents the mass of the fragment and parent ions and eV is the accelerating voltage.

In magnetic-sector instruments, metastable ions are normally observed as small broad peaks. However, in GC/MS the analyst looks only at centrioded (processed) data; thus, metastable peaks are not obvious and generally appear as part of the background. Metastable ions, when observed, can be used to link specific product and precursor ions.

Multiply Charged Ions

Certain compounds will not only produce singly charged ions, but also ions with more than one positive charge. In ei, doubly charged ions are most often observed when analyzing a compound that has an aromatic ring structure. Note that some ionization methods such as electrospray ionization will produce a distribution of molecular ions that differ in the number of charges. For doubly charged ions, the observed peak will appear at a m/z value that is one-half the mass of the ion. These ions will have isotope ions that differ in mass not by 1 Dalton, as with singly charged ions, but by 0.5 Daltons. Thus, the ^{13}C isotope peak for a doubly charged ion with a mass of 602 Daltons will appear at a m/z 301.5.

The Analytic Power of Mass Spectrometry

Although mass spectrometers do nothing more than measure the mass and abundance of ions, they are very powerful analytic instruments. Part of the reason for the power of this technique is its extremely high sensitivity. A complete mass spectrum can be obtained on a few nanograms of material (sometimes less), and selected ions can be observed consuming only a few picograms. The ability to obtain the molecular weight and characteristic fragment ions is frequently sufficient to identify materials without help from other analytic methods. Combining MS with separation methods such as GC has produced an extremely powerful analytic method by which materials present in complex matrices at low ppm levels can be identified, and ppb levels can be verified. The addition of retention time, accurate mass determination, and CID of selected ions further increases the power of GC/MS for the identification of unknown compounds.

References

1. James, A. T., and Martin, A. J. P. Gas liquid partition chromatography. *Biochem. J. Proc.*, *50*, 1952.
2. Grob, K. *Classical Split and Splitless Injections in Capillary GC*. London: Huethig Publishing LTD., 1988.
3. For a discussion of the sulfur chemiluminescence detector, see Shearer, R. L. *American Laboratory*, *12*, 24, 1994.
4. Thomson, J. J. *Rays of Positive Electricity and Their Applications to Chemical Analysis*. Logmaus, 1913.
5. For a more in-depth discussion, see Dass, C. Chapter 1 in D. M. Desiderio, Ed. *Mass Spectrometry: Clinical and Biomedical Applications* (Vol. 2). New York: Plenum Press, 1994.
6. Busch, K. L., Glish, G. L., and McLuckey, S. A. *Mass Spectrometry: Techniques and Applications of Tandem Mass Spectrometry*. New York: VCH Publishers, 1988.
7. For further reading on chemical ionization mass spectrometry, see Harrison, A. G. *Chemical Ionization Mass Spectrometry*. Boca Raton, FL: CRC Press, 1983.
8. For a comprehensive treatment of metastable ions, see Cooks, R. G., Beynon, J. H., Caprioli, R. M., and Lester, G. R. *Metastable Ions*. New York: Elsevier Scientific, 1973.

C h a p t e r 2

Interpretation of
Mass Spectra

There are many ways to interpret mass spectra. Frequently, prior knowledge or the results from a library search dictate the method. The proceeding is a brief description of an approach to mass spectral interpretation that is especially useful when little is known about the compounds in the sample.

Locate or Deduce Molecular Ion

Examine the high-mass region for a possible molecular ion (M). The molecular ion may not be present but can be deduced by considering that ions with these masses can occur: M + 1, M, M − 1, M − 15, M − 18, M − 19, M − 20, M − 26, M − 27, M − 28, M − 29, M − 30, M − 31, M − 32, M − 34, M − 35, M − 40, M − 41, M − 42, M − 43, M − 44, M − 45, M − 46, and so on. Successful interpretation depends on identifying or deducing the molecular ion. The familiar Nitrogen Rule is helpful in eliminating impossible molecular ion candidates. *The Nitrogen Rule states that the observed molecular ion in electron impact ionization is of even mass if the unknown contains either an even number of or no nitrogen atoms. A molecular ion with an odd mass contains an odd number of nitrogen atoms.*

The molecular ion must contain the highest number of atoms of each element present. For example, if a lower-mass ion contains four chlorine atoms, but the highest-mass ion observed contains only three, then at least 35 Daltons should be added to the highest-mass ion observed to deduce the molecular ion.

An abundant molecular ion may indicate that an aromatic compound or highly unsaturated ring compound is present. If no molecular ion is observed and one cannot be deduced, the use of chemical ionization (ci), negative chemical ionization (nci), fast atom bombardment (FAB), or electrospray ionization (ESI) should provide a molecular ion.

Preparation of Trimethylsilyl Derivative

Another way to establish the molecular weight is by preparing derivatives of hydroxyl, amino, or carboxylic acid groups. After preparing the trimethylsilyl (TMS) derivative and obtaining a spectrum of the sample, it is possible to discover which GC peak(s) contain the TMS derivatives by plotting the reconstructed ion chromatogram for m/z 73. The molecular weight of the TMS derivative is determined from the M − 15 peak, which should be a prominent high-mass ion in the spectrum. If two high-mass peaks separated by 15 mass units are observed, then the highest-mass peak is usually the molecular weight of the TMS derivative. If a high-mass peak of odd mass is observed and a peak 15 mass units above is absent, then the molecular weight of the TMS derivative is the mass of the odd mass peak plus 15 mass units (for monosaccharides it may be as much as 105 mass units above the highest mass peak). To determine the original molecular weight, subtract 72 (the mass of C_3H_8Si) mass units for each active hydrogen present. (See Appendix 1 for the method to determine active hydrogens in a single GC/MS run using TMS derivatives.)

Select Structural Type

Characteristic Fragment Ions

Assign possible structures to all abundant fragment ions from the tabulated ions listed in the structurally significant tables of Part III. Two or more ions together may define the type of compound. For example, the presence of the following ions suggest specific compounds:

m/z 15, 29, 43, 57, 71, 85, 99: Aliphatic hydrocarbons
m/z 19, 31, 50, 69: Perfluoro compounds
m/z 30, 44, 58: Amines
m/z 31, 45, 59: Alcohols or ethers
m/z 39, 50, 51, 52, 63, 65, 76, 77, 91: Aromatic hydrocarbons
m/z 41, 54, 68: Aliphatic nitriles
m/z 41, 55, 69: Unsaturated hydrocarbons
m/z 41, 69: Methacrylates

m/z   43, 58: Methyl ketones
m/z   43, 87: Glycol diacetates
m/z   44, 42: $(CH_3)_2N-$
m/z   53, 80: Pyrrole derivatives
m/z   55, 99: Glycol diacrylates
m/z   59, 72: Amides
m/z   60, 73: Underivatized acids
m/z   61, 89: Sulfur compounds
m/z   67, 81, 95: 1-Acetylenes
m/z   69, 41, 86: "Segmented" fluoromethacrylates
m/z   69, 77, 65: "Segmented" fluoroiodides
m/z   74, 87: Methyl esters
m/z   76, 42, 61: $(CH_3)_2NS-$
m/z   82, 67: Cyclohexyl compounds
m/z   83, 82, 54: a cyclohexyl ring
m/z   87, 43: Glycol diacetates
m/z   89, 61: Sulfur-containing compounds
m/z   86, 100, 114: Diamines
m/z   99, 55: Glycol acrylates
m/z   104, 91: Alkylbenzenes
m/z   104, 117: Alkylbenzenes

Use Part III of this book for further correlations.

Identify Fragments Lost from the Molecular Ion

Examine fragment ions to determine the mass of the neutral fragments that were lost from the molecular ion, even though these high-mass peaks may be of low abundances. Compare the neutral loss from the molecular ion with the neutral losses tabulated in Part III to see if these losses agree with the suspected structural type.

Examine the Library Search

Even though a good fit is not obtained, the library search may indicate the structural type. Review the characteristic fragment pathways of the suspected structural type in Part II of this book, and check Part III to determine if the ions observed and neutral losses correspond to the suggested structural type.

Table 2.1.   Isotopic distributions of some elements found in organic compounds

Most abundant isotope (%)	Isotope	M + 1 (%)	M + 2 (%)
$^{12}C = 100$	^{13}C	1.1	—
$^{32}S = 100$	^{34}S	0.8 (^{33}S)	4.4
$^{35}Cl = 100$	^{37}Cl	—	32
$^{79}Br = 100$	^{81}Br	—	98

List Probable Molecular Formula with Calculated Rings Plus Double Bonds

Determination of the Probable Molecular Formula

Examine the low-mass fragment ions. The low-mass ions (which may be present in low abundances) indicate which elements to consider when determining the molecular formula as well as the compound class. Assign possible structures to all abundant fragment ions from the tabulated ions in the "structurally significant" tables of Part III. Two or more ions together may define the type of compound. For example, ions at masses 31, 45, and 59 suggest oxygen-containing compounds, such as alcohols, ethers, ketones, and so forth. Mass 51, in the absence of masses 39, 52, 63, and 65, indicates the presence of carbon, hydrogen, and fluorine. Masses 31, 50, and 69 imply that fluorocarbons are present. Mass 47, in the absence of chlorine, suggests a compound containing carbon, oxygen, and fluorine. This information is also helpful in determining what elements to consider when using accurate mass measurement data.

Look for characteristic isotopic abundances that show the presence of bromine, chlorine, sulfur, silicon, and so on. If the deduced molecular ion is of sufficient intensity, the probable molecular formula may be determined using the observed isotopic abundances of the molecular ion region. Set the deduced molecular ion, M, at 100% abundance, and then calculate the relative abundances of M + 1 and M + 2 either manually or using the data system.

Table 2.1 may be useful for calculating the number of carbon, bromine, chlorine, and sulfur atoms in the molecular formula. This table shows that for every 100 ^{12}C atoms there are 1.1 ^{13}C atoms. Also, for every 100 ^{32}S atoms, there are 0.8 ^{33}S atoms and 4.4 ^{34}S atoms. The following examples demonstrate how the molecular formula can be deduced from the isotope abundances of the molecular ion.

Example 2.1

Observed isotopic abundances for the molecular ion region:

m/z 78 = 100% [M]
        79 = 6.6% $[M + 1]^+$

$\dfrac{6.6}{1.1}$ = 6 carbon atoms (Divide ^{13}C isotope abundance from Table 2.1
   into observed $[M + 1]^+$ abundance.)
6 × 12 = 72 (Multiply 6 carbon atoms by monoisotopic mass of
   carbon.)
78 − 72 = 6 hydrogen atoms (Subtract 72 from observed molecular
   ion.)
The maximum carbon atoms are 6, and the maximum hydrogen atoms
   are 6. The probable molecular formula is C_6H_6.

Example 2.2

Observed isotopic abundances for the molecular ion region:

m/z 100 = 12.2% $[M]^+$
        101 = 1.0% $[M + 1]^+$

Setting M to 100% abundance gives m/z 100 = 100% $[M]^+$
                                          101 = 8.2% $[M + 1]^+$

$\dfrac{8.2}{1.1}$ = 7 or 8 carbon atoms
For 7 carbon atoms, the maximum number of hydrogen atoms is
   100 − (7 × 12) = 16.
For 8 carbon atoms, the maximum number of hydrogen atoms is
   100 − (8 × 12) = 4.
Obviously, C_7H_{16} is a more probable molecular formula than C_8H_4.

As noticed by this example, as the molecular ion becomes smaller, the accuracy of the method decreases and is unusable if the M + 1 and/or M + 2 ions are not observed or the elements present are not known.

Example 2.3

Observed isotopic abundances for the molecular ion region:

m/z 146 = 100% $[M]^+$
    147 = 6.6% $[M + 1]^+$
    148 = 64% $[M + 2]^+$

From the isotope abundances listed in Table 2.1, it is obvious that the M + 2 ion abundance in this example is due to two chlorine atoms.

$\dfrac{64}{32}$ = 2 chlorine atoms $\dfrac{6.6}{1.1}$ = 6 carbon atoms

$2 \times 35 =$ 70   (2 chlorine atoms have a mass of 70.)
$6 \times 12 =$ _72_   (6 carbon atoms have a mass of 72.)
          142   (70 + 72 = 142)
m/z 146 − m/z 142 = 4 hydrogen atoms. (Then, the molecular formula is $C_6H_4Cl_2$.)

This method can sometimes be used for determining the probable elemental composition of fragment ions. However, it is not as generally applicable and does not replace accurate mass measurement for determining molecular formulae and elemental compositions.

To calculate M + 1 and M + 2 of molecules containing only C, H, O, and N, the following may be used:

M + 1 = 1.1 (no. of carbons) + 0.37 (no. of nitrogens) + 0.04 (no. of oxygens)

$$M + 2 = \frac{[1.1 \times \text{no. of carbons}]^2}{200} + 0.2 \text{ (no. of oxygens)}$$

For example, the ^{13}C isotope will contribute approximately 2% to the M + 2 in compounds containing 20 carbon atoms.

Rings Plus Double Bond Calculation

The number calculated for rings (R) plus double bonds (DB) must be either zero or a whole number and agree with the suspected structural type.

R + DB = (no. of carbons) − 1/2 (Hydrogens + Halogens) + 1/2 (Nitrogens) + 1

The following are examples of calculating rings plus double bonds.

Example 2.4

$C_{12}H_{10}N_2$
R + DB = 12 − 1/2(10) + 1/2(2) + 1 = 9

Example 2.5

$C_{13}H_{10}O$
R + DB = 13 − 1/2(10) + 1/2(0) + 1 = 9

The unknown structures for the previous examples have nine double bonds and/or rings.

In the R + DB formula, you may make the following substitutions:

For	Substitute
O	CH_2
N	CH
Halogens	CH_3
S	CH_2 (Valence of 2)
S	C (Valence of 4)
Si	(Treat as a carbon)

It is now necessary to know that a saturated hydrocarbon has the formula C_nH_{2n+2}. Therefore, a compound having a formula $C_{12}H_{10}N_2$ would be equivalent to $C_{12}H_{10}$ + 2(CH) or $C_{14}H_{12}$. A saturated C_{14} hydrocarbon would have the formula $C_{14}H_{30}$. $C_{14}H_{12}$ has 18 fewer hydrogens than the saturated C_{14} hydrocarbon, therefore, there are $\frac{18}{2}$ or 9 rings and double bonds.

List Structural Possibilities

List possible structures or partial structures consistent with the mass spectral data. If probable molecular formulae are tabulated and the structure is still unknown, possible structures can easily be obtained from such sources as *Beilstein, Merck Index, Handbook of Chemistry, Handbook of Chemistry and Physics*, and even chemical catalogs such as the *Aldrich Catalog*, and so on. Check to see that the calculated rings plus double bonds agree with the suspected structure type.

For example, from the *Merck Index*:

$C_{12}H_{10}N_2$, [structure: benzene ring—N=N—benzene ring]

Azobenzene
R + DB = 9

$C_{13}H_{10}O_1$, [structure: benzene ring—C(=O)—benzene ring]

Benzophenone
R + DB = 9

To distinguish between azobenzene and benzophenone, assuming reference spectra are not available for these compounds, it is a good idea to examine the mass spectra of aromatic ketones, such as acetophenone, butyrophenone, diphenyldiketone, and so forth. Complete identification is assured by obtaining or synthesizing the suspected component and analyzing it on the GC/MS system under the same GC conditions. If the retention time and the mass spectrum agree, then the identification is confirmed.

Hint: To distinguish these compounds without elemental composition or standards for GC retention time, split the GC effluent to a FID, a nitrogen-phosphorus detector, and the mass spectrometer, simultaneously. Using this splitter system, it is easy to determine if the GC peak contains nitrogen. Also, the analyst can differentiate between azobenzene and benzophenone by using the methoxime derivative.

Compare the Predicted Mass Spectra of the Postulated Structures with the Unknown Mass Spectrum

After the possible structures are obtained, predict their mass spectra by examining the mass spectra of similar structures. Also, the GC retention time may eliminate certain structures or isomers. Discuss these results with the originator of the sample to determine the most probable structure. With experience, it is usually possible to determine which fragment peaks are reasonable for a given type of structure.

Compound Identification Examples

Example 2.6

The unknown gave a molecular ion at m/z 193 with fragment ions at m/zs 174, 148, and 42. From the abundance of the molecular ion, it is probably aromatic, and according to the Nitrogen Rule, contains at least one nitrogen atom. From accurate mass measurement data and an examination of the isotopic abundances in the molecular ion region, the molecular formula was found to be $C_{11}H_{15}NO_2$.

$$R + DB = 11 - 1/2(15) + 1/2(1) + 1 = 5$$

		m/z
$C_{11}H_{15}NO_2$	M	193
$C_9H_{10}NO_2$	$M-C_2H_5$	174
$C_9H_{10}NO$	$M-OC_2H_5$	148

These losses suggest an ethyl ester. Looking up m/z 148 in Part III suggests:

$$(CH_3)_2N-\underset{}{\bigcirc}-\overset{\displaystyle O}{\underset{\displaystyle \|}{C}}-$$

Examination of the low-mass region showed a m/z-42 peak that is characteristic of $(CH_3)_2N-$. The proposed structure was $(CH_3)_2N-C_6H_4-CO_2C_2H_5$.

Example 2.7

The mass spectrum of the unknown compound showed a molecular ion at m/z 246 with an isotope pattern indicating that one chlorine atom and possibly a sulfur atom are present. The fragment ion at m/z 218 also showed the presence of chlorine and sulfur. The accurate mass measurement showed the molecular formula to be $C_{13}H_7OSCl$; $R + DB = 10$.

m/z 246 $C_{13}H_7OSCl$ M$^{+\cdot}$
     218 $C_{12}H_7SCl$ M$-CO$

The loss of CO suggested a cyclic ketone. Looking up possible structures in Part III, a chlorothioxanthone structure was indicated.

Example 2.8

Nuclear magnetic resonance (NMR) and infrared spectroscopy (IR) narrowed an unknown down to two possible structures:

$$I \qquad\qquad + \qquad\qquad II$$

GC/MS was used to distinguish between the two structures. The mass spectrum showed a molecular ion at m/z 260. The fragment ions occurred at m/z 245, 241, 231, and 205. This is a good example of nitrogen atom-influenced fragmentation; therefore, structure I was highly favored.

C h a p t e r   3

Quantitative GC/MS

It is crucial in quantitative GC to obtain a good separation of the components of interest. Although this is not critical when a mass spectrometer is used as the detector (because ions for identification can be mass selected), it is nevertheless good practice. If the GC effluent is split between the mass spectrometer and FID detector, either detector can be used for quantitation. Because the response for any individual compound will differ, it is necessary to obtain relative response factors for those compounds for which quantitation is needed. Care should be taken to prevent contamination of the sample with the reference standards. This is a major source of error in trace quantitative analysis. To prevent such contamination, a method blank should be run, following all steps in the method of preparation of a sample except the addition of the sample. To ensure that there is no contamination or carryover in the GC column or the ion source, the method blank should be run prior to each sample.

Peak Area Method

A rough estimate of the concentrations of components in a mixture can be obtained using peak areas. This method assumes that the area percent is approximately the weight percent. The area of each peak is divided by the sum of the areas of all peaks,

$$\text{Component } X\% = \frac{A_x(100)}{\sum_{n=1}^{n_i=n} A_{n_i}} \quad (1)$$

where A_x = peak area = $H \times W$ (H = peak height, W = peak width at half height). Most data systems are able to calculate peak areas. n is the total number of peaks in the chromatogram that are summed to give the area of all peaks.

This result can be used to prepare a synthetic mixture to obtain relative response factors.

Relative Response Factor Method

Using the peak area method, prepare a standard solution in which the amounts of each component will approximate the amounts found in the sample being analyzed. From the standard solution, obtain the GC peak areas for each component. Assign to one of the major components a relative response factor (RF) of 1.0. This component is the reference. The response factors for the other components are obtained in the following manner.

$$RF_x = \frac{A_R \cdot W_x}{A_x \cdot W_R} = \frac{SA_R}{SA_x} \quad (2)$$

Where SA_R is the specific area of the reference peak, and SA_x is the specific area of component x. A_R is the GC peak area of the reference, A_x is the GC peak area of component x, W_R is the weight of the reference, and W_x is the weight of component x. The weight percent of component x can be obtained from the sample chromatogram by using the relative response factors in the following equation:

$$W_x\% = \frac{A_x \cdot RF_x \cdot 100}{\sum (A_n \cdot RF_n)} \quad (3)$$

$\Sigma(A_n \cdot RF_n)$ is the sum of the areas times the individual response factors for all the peaks in the chromatogram. The amount injected should be the same for both the standard and the unknown.

An example of this method follows:

Component	Weight	Peak Area
n-butyl alcohol (1)	0.3904 g	242.65
cyclohexanone (2)	0.4840 g	338.01
N-methylpyrrolidone (3)	0.6412 g	268.66
N,N-dimethylformamide (4)	0.7533 g	174.29

RF(1) = (174.29 × 0.3904)/(242.65 × 0.7533) = 0.372

RF(2) = (174.29 × 0.4840)/(338.01 × 0.7533) = 0.331

RF(3) = (174.29 × 0.6412)/(268.66 × 0.7533) = 0.552

RF(4) = 1.000 (by definition)

Component	$\Sigma(A_x \cdot RFx)$
(1)	90.31
(2)	111.98
(3)	148.36
(4)	174.29

Thus, $\Sigma(A_n \cdot RF_n)$ = 524.94, and W(1)% = 90.31 × 100/524.94 = 17.2%.

Internal Standard Method

Accurate quantitation in GC/MS requires the addition of a known quantity of an internal standard to an accurately weighed aliquot of the mixture (matrix) being analyzed. The internal standard corrects for losses during subsequent separation and concentration steps and provides a known amount of material to measure against the compound of interest. The best internal standard is one that is chemically similar to the compound to be measured, but that elutes in an empty space in the chromatogram. With MS, it is possible to work with isotopically labeled standards that co-elute with the component of interest, but are distinguished by the mass spectrometer.

A known weight of the internal standard (W_{is}) is added to the sample matrix (which has been carefully weighed) in an amount that is close to the amount expected for the compound being measured (A_x). The same quantity of internal standard is added to several vials containing known weights of the compound(s) for which quantitation is needed (W_n). The solvent should be the same for both the known solutions and the unknown to be measured. For accurate work, the concentrations in the vials should bracket the concentration of the sample of interest. A blank should be prepared by adding the internal standard to the solvent used for the sample. This is especially important with trace analysis to prevent contamination during handling.

Once the internal standard has been added to the unknown matrix and is thoroughly mixed, the solution can be concentrated, if necessary, for GC/MS analysis. Chemical derivatization is also performed if needed. Normally, samples and reference standards are evaporated to dryness and then redissolved in a carefully measured quantity of solvent. For trace samples, se-

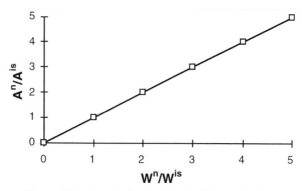

Figure 3.1. Internal standard method for quantitation.

lected ion monitoring of the sample of interest and the internal standard is required.

At this point, the solution containing the component to be measured (A_x) also contains any other compounds from the original matrix that are soluble in the solvent used in the analysis. For the analysis to be accurate, other components in the matrix cannot interfere by eluting at the same retention time as the components to be measured. For accurate MS analyses, the matrix component must not interfere with production of the ions being measured for either the internal standard or the component to be measured. In some cases, to eliminate interferences, it may be necessary to increase the resolution of the mass spectrometer by narrowing the mass window being monitored. Alternatively, MS/MS can be used to avoid chemical interference (see Chapter 1).

The peak area of the unknown (A_x) relative to the peak area of the internal standard (A_{is}) is obtained. Conversion of the measured ratio to a concentration is achieved by comparing it to area ratios of the solutions of known analyte concentration, to which the same quantity of internal standard has been added. A graph of the ratio of the peak area of the component to be measured (A_n) to the peak area of the internal standard (A_{is}) versus the ratio of the weight of the component to be measured (W_n) to the weight of the internal standard (W_{is}) for the known solutions results in a graph from which the concentration of the component(s) in the unknown matrix (A_x) can be determined (Figure 3.1).

Making Standard Solutions

The following procedure may be useful in making standard solutions:

- Accurately weigh *ca.* 10 mg of the standard using an analytic balance.
- Quantitatively transfer this amount to a clean, dry 100-ml volumetric flask. Fill the flask to the mark with solvent.

- Shake the volumetric flask vigorously to allow the material to dissolve and ensure homogeneity.

This 100-ml volumetric flask contains an accurately known concentration that is close to 100 ppm = 100 μg/ml = 100 ng/μl. Label the flask using the concentration determined from the actual weights.

- Accurately pipet 1.00 ml to a second 100-ml volumetric flask.
- Fill this flask to the mark with solvent. Shake the flask vigorously to ensure homogeneity.

This flask contains *ca.* 1 ppm = 1 μl/ml = 1 ng/μl. Label the flask with the actual concentration.

Part I References

Useful GC/MS Books

Blau, K., and Halket, J., Eds., *Handbook of Derivatives for Chromatography*. New York: Wiley, 1993.

Gudzinowicz, B. J., Gudzinowicz, M. J., and Martin, H. F. *Fundamentals of Integrated GC-MS* (Vols. 1–3). New York: Marcel Dekker, 1976.

Karasek, F. W., and Clement, R. E. *Basic Gas Chromatography-Mass Spectrometry—Principles and Techniques*. New York: Elsevier, 1991.

Linskens, H. F., and Jackson, J. F., Eds. *Gas Chromatography/Mass Spectrometry* (Vol. 3). New York: Springer, 1986.

Message, G. M. *Practical Aspects of Gas Chromatography/Mass Spectrometry*. New York: Wiley & Sons, 1984.

The following books are now out of print, but if available through your library or a private collection, are a valuable resource.

Beynon, J. H., Saunders, R. A., and Williams, A. E. *The Mass Spectra of Organic Compounds*. New York: Elsevier, 1968.

Budzikiewicz, H., Djerassi, C., and Williams, D. H. *Interpretation of Mass Spectra of Organic Compounds.* San Francisco: Holden-Day, 1964.

Domsky, I. I., and Perry, J. A. *Recent Advances in Gas Chromatography*. New York: Marcel Dekker, 1971.

Hamming, M. C., and Foster, N. G. *Interpretation of Mass Spectra of Organic Compounds.* San Diego: Academic Press, 1972.

McFadden, W. H. *Techniques of Combined Gas Chromatography/Mass Spectrometry: Applications in Organic Chemistry*. New York: Wiley & Sons, 1973.

Porter, Q. N., and Baldas, J. *Mass Spectrometry of Hetrocyclic Compounds*. New York: Wiley-Interscience, 1971.

Watson, J. T. *Introduction to Mass Spectrometry: Biomedical, Environmental & Forensic Application*. New York: Raven Press, 1975.

Williams, D. H., and Howe, I. *Principles of Organic Mass Spectrometry.* New York: McGraw-Hill, 1972.

Gas Chromatography

ASTM Test Methods from ASTM. ASTM, 1916 Race Street, Philadelphia, PA 19103.

Eiceman, G. A., Hill, H. H., Jr., and Davani, B. Gas chromatography detection methods, *Anal. Chem.,* 66, 621R, 1994.

Grob, R. L. *Modern Practice of Gas Chromatography.* New York: Wiley-Interscience, 1985.

Heijmans, H., de Zeeuw, J., Buyten, J., Peene, J., and Mohnke, M. PLOT columns. *American Laboratory, August,* 28C, 1994.

Johnson, D., Quimby, B., and Sullivan, J. Atomic emission detector for GC. *American Laboratory, October,* 13, 1995.

Katritzky, A. R., Ignatchenko, E. S., Barcock, R. A., and Lobanov, V. S. GC retention times and response factors. *Anal. Chem.,* 66, 1799, 1994.

Klemp, M. A., Akard, M. L., and Sacks, R. D. Cryofocusing sample injection method. *Anal. Chem.,* 65, 2516, 1993.

McNair, H. M. Gas chromatography. *LC-GC,* 11(11), 794, 1993.

Mass Spectrometry

Busch, K. L. Getting a charge out of mass spectrometry. *Spectroscopy,* 9, 12, 1994.

Russell, D. H., Ed. *Experimental Mass Spectrometry.* New York: Plenum Press 1994.

Watson, J. T. *Introduction to Mass Spectrometry.* New York: Raven Press, 1985.

What is Mass Spectrometry? The American Society for Mass Spectrometry (ASMS), 815 Don Gaspar Drive, Santa Fe, NM 87501.

MS Instrumentation

Dass, C. Chapter 1. In D. M. Desiderio, Ed. *Mass Spectrometry: Clinical and Biomedical Application* (Vol. 2). New York: Plenum Press, 1994.

Chemical Ionization

Harrison, A. G. *Chemical Ionization Mass Spectrometry* (2nd ed.). Boca Raton, FL: CRC Press, 1992.

Mass Spectral Interpretation

Ardrey, R. E., Allen, A. R., Bal, T. S., and Moffat, A. C. *Pharmaceutical Mass Spectra.* London: Pharmaceutical Press, 1985.

Eight Peak Index of Mass Spectra (4th ed.). Boca Raton, FL: CRC Press, 1992.

Hites, R. A. *Handbook of Mass Spectra of Environmental Contaminants.* Boca Raton, FL: CRC Press, 1992.

McLafferty, F. W., and Stauffer, D. B. *Important Peak Index of the Registry of Mass Spectral Data.* New York: Wiley-Interscience, 1991.

McLafferty, F. W., and Turecek, F. *Interpretation of Mass Spectra* (4th ed.). Mill Valley, CA: University Science Books, 1993.

McLafferty, F. W., and Venkataraghavan. *Mass Spectral Correlations.* San Diego: ACS, 1982.

Pfleger, K., Maurer, H., and Weber, A. *Mass Spectral and GC Data of Drugs, Poisons, and Their Metabolites.* New York: VCH Publishers, 1992.

Sunshine, I. *Handbook of Mass Spectra of Drugs.* Boca Raton, FL: CRC Press, 1981.

Quantitation by GC/MS

Bergner, E. A., and Lee, W.-N. P. Testing GC/MS systems for linear response. *J. Mass Spectrom., 10,* 778, 1995.

Cai, Z., Sadagopa, R. V. M., Giblin, D. E., and Gross, M. L. GC/HRMS. *Anal. Chem., 65,* 21, 1993.

Girault, J., Longueville, D., Ntzanis, L., Couffin, S., and Fourtillan, J. B. Quantitation of urine using GC/negative ion CI MS. *Biol. Mass Spectrom., 23,* 572, 1994.

Hayes, M. J., Khemani, L., and Powell, M. L. Quantitation using capillary GC/MS. *Biol. Mass Spectrom., 23,* 555, 1994.

Herman, F. L. Tandem detectors to quantitate overlapping GC peaks. *Anal. Chem., 65,* 1023, 1993.

Watson, J. T., Hubbard, W. C., Sweetman, B. J., and Pelster, D. R. Quantitative analysis of prostaglandins by selected ion monitoring GC/MS. *Advan. in Mass Spectr. in Biochem. & Med II.* New York: Spectrum Publications, 1976.

Review Articles

Grayson, M. A. GC/MS. *J. Chromatographic Sci., 24,* 529, 1986.

Guiochon, G., and Guillemin, C. L. Gas chromatography, *Rev. Sci. Instrum., 61,* 3317, 1990.

Mikaya, A. I., and Zaikin, V. G. Reaction gas chromatography/mass spectrometry. *Mass Spectrom. Rev., 9,* 115, 1990.

Part II

GC Conditions, Derivatization, and Mass Spectral Interpretation of Specific Compound Types

C h a p t e r 4

Acids

I. GC Separations of Underivatized Carboxylic Acids

A. Aliphatic Acids

  1. Capillary columns

    a. C_1–C_5:   30 m DB-FFAP or HP-FFAP column at 135°,
      C_2–C_{10}^+:   30 m DB-FFAP column (or equivalent), 50–240° at 10°/min, run for 1 hr (approximately 100 ppm can be detected).

    b. C_2–C_7:   25–30 m OV-351 column at 145°.

    c. C_1–C_7:   30 m DB-WAX column, 80–230° at 10°/min.

  2. Packed columns

    a. C_2–C_5:   2 m SP-1200 column + H_3PO_4 or SP-1200 column + H_3PO_4 at 125°.

    b. C_1–C_{10}:   2 m SP-1220 column + H_3PO_4, 70–170° at 4°/min.

    c. C_1–C_9:   2 m Chromosorb 101, Porapak Q, or Porapak QS column can be used.

B. Simple Mixtures (which include free acids)

1. Formic acid, acetic acid, and propionic acid
   2 m Porapak QS at 170°.

2. Acetaldehyde, ethyl formate, ethyl acetate, acetic anhydride, and acetic acid
   25 m CP-WAX 52CB column, 50–200° at 5°/min.

C. Aromatic Carboxylic Acids (See the following derivatization procedures.)

II. General Derivatization Procedure for C_8–C_{24} Carboxylic Acids

A. Aliphatic Acids—TMS Derivatives
   For low molecular weight aliphatic acids, try TMSDEA reagent. Otherwise, use MSTFA, BSTFA, or TRI-SIL BSA (Formula P). For analysis of the keto acids, methoxime derivatives should be prepared first, followed by the preparation of the trimethylsilane (TMS) derivatives using BSTFA reagent. This results in the methoxime-TMS derivatives.

III. GC Separation of Derivatized Carboxylic Acids

A. Krebs Cycle Acids

1. Derivatives: Krebs cycle acids have been analyzed using only the TMS derivatives, even though some are keto acids.

2. GC conditions: 30 m DB-1 column, 60–250° at 5°/min.

Acid*	MW of TMS Derivatives
Malic	350
Fumaric	260
Succinic	262
2-Ketoglutaric	290
Oxalsuccinic	406
Isocitric	480
cis-Aconitic	390
Citric	480

*Author has not separated all these acids as mixtures.

B. α-Keto Acids—Methoxime-TMS Derivatives

1. Derivatives: Add 0.25 ml of methoxime hydrochloride in pyridine and let stand at room temperature for 2 hr. Evaporate to dryness with clean, dry nitrogen. Add 0.25 ml of BSTFA, MSTFA, or BSA reagent and let stand for 2 hr at room temperature.

2. GC conditions: 30 m DB-1 column, 60 (2 min)–200° at 10°/min–250° at 15°/min.

3. The following components are in the order of elution using the GC conditions given previously.

   a. *Pyruvic acid-MO-TMS*
   $CH_3C(NOCH_3)C(O)OTMS$
   Major ions: *m/z* 174.0586, 115
   Highest mass ion observed: *m/z* 189.0821

   b. *α-Ketobutyric acid-MO-TMS*
   $CH_3CH_2C(NOCH_3)C(O)OTMS$
   Major ions: *m/z* 73, 89
   For selected ion monitoring (SIM), plot 188.0742 (M − CH_3)

   c. *α-Ketoisovaleric acid-MO-TMS*
   $(CH_3)_2CHC(NOCH_3)C(O)OTMS$
   Major ions: *m/z* 73, 89, 100
   For SIM, plot 186.0948 (M − OCH_3)
   Highest mass ion observed: *m/z* 202 or 217

   d. *α-Keto-β-methylvaleric acid-MO-TMS*
   $CH_3CH_2CH(CH_3)C(NOCH_3)C(O)OTMS$
   Major ions: *m/z* 73, 89
   For SIM, plot 200.1107 (M − OCH_3)
   Highest mass ion observed: *m/z* 216
   m/z 203 distinguishes this isomer from the following component. Both isomers have *m/z* 189, 200, and 216.

   e. *α-Ketoisocaproic acid-MO-TMS*
   $(CH_3)_2CHCH_2C(NOCH_3)C(O)OTMS$
   Major ions: *m/z* 189, 200, 216
   For SIM, plot 216.1056

f. *2,3-Dihydroxyisovaleric acid-TMS*

$$(CH_3)_2\underset{\underset{\displaystyle OTMS}{|}}{C}\text{------}\underset{\underset{\displaystyle OTMS}{|}}{C}HC(O)OTMS$$

Most abundant ion: *m/z* 131
For SIM, plot 292.1346

g. *α-Isopropylmaleic acid-TMS*

$$\overset{\overset{\displaystyle C(O)OTMS}{|}}{(CH_3)_2CHC}\!=\!CHC(O)OTMS$$

For SIM, plot 287.1135 (M − CH$_3$)

h. *α, β-Dihydroxy-β-methylvaleric acid-TMS*

$$CH_3CH_2\!-\!\underset{\underset{\displaystyle OTMS}{|}}{\overset{\overset{\displaystyle CH_3}{|}}{C}}\text{------}\underset{\underset{\displaystyle OTMS}{|}}{C}H\!-\!C(O)OTMS$$

Abundant ion: *m/z* 145
For SIM, plot 292.1346

i. *α-Isopropylmalic acid-TMS*

$$(CH_3)_2CH\!-\!\underset{\underset{\displaystyle C(O)OTMS}{|}}{\overset{\overset{\displaystyle OTMS}{|}}{C}}CH_2C(O)OTMS$$

Major ions: *m/z* 275, 261, 349
For SIM, plot 275.1499 (M − C(O)OTMS)

j. *β-Isopropylmalic acid-TMS*
Note: The α-isomer elutes slightly ahead of the β-isomer.

$$(CH_3)_2CHC\underset{\underset{\displaystyle TMSO-CH-C(O)OTMS}{|}}{H}\!-\!C(O)OTMS$$

Major ions: m/z 275, 191, 231, 305
For SIM, plot 275.1499 (M − C(O)OTMS)

C. Itaconic Acid, Citraconic Acid, and Mesaconic Acid

1. Derivatives: Add 0.25 ml of MTBSTFA reagent to less than 1 mg of sample and heat at 60° for 30 min.

2. GC conditions: 30 m DB-210 column, 60–220° at 10°/min.

D. Higher-Boiling Acids Such as Benzoic and Phenylacetic Acids

1. Derivatives: Add 0.25 ml of MSTFA or TRI-SIL BSA (Formula P) to the dried extract and heat at 60° for 30 min.

2. GC conditions: 25 m CPSIL-5 column, 100–210° at 4°/min.

E. Organic Acids in Urine

1. Derivatives: 1 ml of urine adjusted to pH 8 with $NaHCO_3$ solution. Add methoxime hydrochloride or ethoxime hydrochloride. Dissolve and mix thoroughly, and then saturate the solution with NaCl. Adjust the solution to pH 1 with 6N HCl.
   Extract with three 1-ml volumes of diethyl ether (top layer) followed by three 1-ml volumes of ethyl acetate. Combine the extractions and evaporate to dryness with clean, dry nitrogen. Add 10 μl of pyridine and 20 μl of BSTFA reagent. Cap the vial and heat at 60° for 7 min.

2. GC conditions: 30 m DB-1 column, 120 (4 min)–290° at 8°/min and hold for 25 min.

3. Acids commonly found in urine: Some of the acids found in urine are given in the proceeding text. We have found as many as 100 GC peaks in urine samples, which include urea and other nonacids. The following components are in order of elution.

Component	MW of TMS Derivatives
Phenol	166
Lactic acid	234
Glycolic acid	220
Oxalic acid	234
Hydroxybutyric acid	262
Benzoic acid	194
Urea	204
Phosphoric acid	314
Phenylacetic acid	208
Succinic acid	262
Glyceric acid	322
Fumaric acid	260
Glutaric acid	276
Capric acid	244
Malic acid	350
Hydroxyphenylacetic acid	296
Pimelic acid	304
Tartaric acid	438
Suberic acid	318
Aconitic acid	390
Hippuric acid	323
Citric acid	480
Isoascorbic acid	464
Indoleacetic acid	319
Gluconic acid	628
Palmitic acid	328
Uric acid	456

F. Bile Acids

1. Derivatives: Acetylated methyl esters are the most suitable derivatives (e.g., deoxycholic acid and cholic acid).

   Evaporate the sample to dryness with clean, dry nitrogen. Add 250 μl of methanol and 50 μl of concentrated sulfuric acid. Heat at 60° for 45 min. Add 250 μl of distilled water and allow to cool. Then add 50 μl of chloroform or methylene chloride. Shake the mixture for 2 min. Remove the bottom layer with a syringe. Evaporate to dryness with clean, dry nitrogen. Acetylate with 50 μl of three parts acetic anhydride and two parts pyridine for 30 min at 60°. Evaporate to dryness with clean, dry nitrogen. Dissolve the residue in 25 μl of ethyl acetate.

2. GC conditions: 25 m DB-1 column 200–290° at 4°/min.

G. C_6–C_{24} Monocarboxylic Acids and Dicarboxylic Acids as Methyl Esters

1. Derivatives

   a. BF$_3$/methanol: Add 1 ml of BF$_3$/methanol reagent to less than 1 mg of the dry extract. Let the reaction mixture stand overnight or heat at 60° for 20 min. Cool in an ice-water bath and add 2 ml of water. Within 5 min, extract twice with 2 ml of methylene chloride. Evaporate the total methylene chloride extracts (if necessary).

   b. Methanol/acid (preferred method for trace analysis): Using a 1- or 2-ml reaction vial, add less than 1 mg of the sample. Add 250 μl of methanol and 50 μl of concentrated sulfuric acid. Cap the vial, shake, and heat at 60° for 45 min. Cool and add 250 μl of distilled water using a syringe. Add 500 μl of chloroform or methylene chloride and shake the mixture for 2 min. Inject a portion of the chloroform layer into the GC.

   c. Diazomethane: To less than 1 mg of the dry extract, add 200 μl of an ethanol-free solution of diazomethane in diethyl ether. (*Caution:* Do not use ground glass fittings when running this reaction.) This solution is stable for several months if stored in small vials (1–10 ml) in the freezer at $-10°$. Evaporate the methylation mixture and dissolve the residue in methanol.

   d. Methyl-8 reagent: Add 0.25 ml of Methyl-8 reagent to less than 1 mg of the dry extract. Cap the vial and heat at 60° for 15 min.

2. GC conditions

   a. C_{14}–C_{22} unsaturated dibasic acids
      30 m DB-23 or CPSIL-88 column, 75–220° at 4°/min.

   b. 30 m DB-WAX column, 60–200° at 4°/min.

H. Bacterial Fatty Acids

1. Derivatives (See Section III,G.)

2. GC conditions

   a. C_8–C_{20} methyl esters of bacterial acids
      30 m DB-1 column, 150 (4 min)–250° at 6°/min.

3. Types of bacterial fatty acids

   a. Saturated, straight chain: $CH_3\text{-}(CH_2)_n\text{-}COOH$

   b. Unsaturated, straight chain: $CH_3\text{-}(CH_2)_n\text{-}CH=CH\text{-}(CH_2)\text{-}COOH$

   c. Branched chain:

      1. Iso: $(CH_3)_2CH(CH_2)_nCOOH$

      2. Anteiso: $C_2H_5CH(CH_3)(CH_2)_nCOOH$

   d. Cyclic: $CH_3-(CH_2)_n-\underset{\diagdown\;\diagup}{CH-CH}-(CH_2)_n-COOH$
      $\underset{CH_2}{}$

   e. Hydroxy:

      1. α (2-OH): $CH_3-(CH_2)_n-\underset{\underset{OH}{|}}{CH}-COOH$

      2. β (3-OH): $CH_3-(CH_2)_n-\underset{\underset{OH}{|}}{CH}-CH_2-COOH$

I. Cyanoacids: $NC(CH_2)_nCOOH$

   1. GC conditions: 30 m DB-1, 100–275° at 10°/min.

IV. Mass Spectral Interpretation

A. Underivatized Carboxylic Acids

Although carboxylic acids are more often analyzed as methyl esters, there are occasions when they are more easily analyzed as free acids, such as in water at the ppm level.

Abundant ions are observed in the mass spectra of straight-chain carboxylic acids at m/z 60 and 73 from n-butanoic to n-octadecanoic acid. The formation of an abundant rearrangement ion at m/z 60 requires a hydrogen in position four of the carbon chain. Most mass spectra of acids are easy to identify with the exception of 2-methylpropanoic acid, which does not have a hydrogen at the C-4 position and cannot undergo the McLafferty

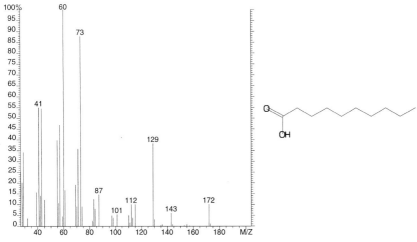

Figure 4.1. Decanoic acid.

rearrangement (see Appendix 10). If the 2-carbon is substituted, instead of m/z 60, the rearrangement ion will appear at m/z 74, 88, and so on, depending on the substitution. Even though methyl esters have a characteristic ion at m/z 74, the mass spectrum of an acid can be distinguished from that of an ester by examining the losses of OH, H_2O, and COOH from the molecular ion of acids in contrast to the loss of OCH_3 in the case of methyl esters. Also, in the higher molecular weight aliphatic acids, the intensities of the molecular ions increase from *n*-butanoic to *n*-octadecanoic acid.

B. Derivatized Carboxylic Acids (See Chapter 12,II.)

C. Mass Spectra of Underivatized Cyano Acids

The molecular ion is usually not observed. Intense ions are observed at m/z 41 and 55. Other characteristic ions are at m/z 60, M − 15, M − 40, M − 46, and M − 59.

D. Sample Mass Spectrum

Examination of the mass spectrum of *n*-decanoic acid (Figure 4.1) shows prominent ions at m/z 60 and 73. A m/z 60 ion (Section III) (see also Appendix 10) suggests the mass spectrum may represent an aliphatic carboxylic acid. This ion in combination with m/z 73

(Section III) is a strong indication of a carboxylic acid. Small peaks at m/z 31 and 45 also suggest the presence of oxygen. The molecular ion appears to be at m/z 172. Subtracting 32 for the two oxygens of the carboxylic acid group leaves 140 Daltons, which is $C_{10}H_{20}$. The compound is decanoic acid.

C h a p t e r   5

Alcohols

I. GC Conditions for Underivatized Alcohols

A. General GC Separations

1. C_1–C_5 alcohols
   2 m Carbowax 1500 on a Carbopak C column, 60–175° at 5°/min, or isothermal at 135°.

2. C_4–C_8 alcohols
   30 m CP-WAX 52CB column, 50–200° at 10°/min.

3. C_8–C_{18} alcohols: 30 m DB-5 column, 50-140° at 10°/min, then 140-250° at 4°/min.

B. Separation Examples

1. Ethanol, 1-propanol, 2-methyl-1-propanol, 2-pentanol, isoamyl alcohol, amyl alcohol
   50 m CP-WAX 52CB column, 60–70° at 2°/min, then 70–200° at 10°/min.

2. Methanol, ethanol, *iso*propyl alcohol, *n*-propyl alcohol, *sec*-butyl alcohol, *n*-butyl alcohol
   30 m GS-Q column, 60-200° at 6°/min.

3. 3-Methyl-1-butanol, 2-methyl-1-butanol
   30 m CP-WAX 51 (or CP-WAX 57CB) column, from 60–175° at 5°/min.

4. Methanol, ethanol, *iso*propylalcohol, *n*-propylalcohol, *tert*-butyl alcohol, 2-butanol, 2-methyl-1-propanol, 1-butanol, 2-pentanol, 2-methyl-1-butanol, 1-pentanol
   30 m Poraplot Q column, 135–200° at 2°/min.

5. 2-Butanol, 1-butanol, 1,3-butanediol, 2,3-butanediol, 1,4-butanediol
   3 m 3% Carbowax 1500 column on 80-200 mesh Carbopack B, 80–225° at 8°/min.

6. Acetaldehyde, methanol, acetone, ethanol, *iso*propyl alcohol, *n*-propyl alcohol
   2 m Carbowax 20M column on Carbopack B at 75°.

II. TMS Derivative of $>C_{10}$ Alcohols

A. Preparation of TMS derivative using less than 1 mg of alcohol add 250 μL MSTFA reagent. Heat at 60° for 5–15 min.

B. GC separation of TMS derivative 30 m DB-5 column, 60–250° at 10°/min.

III. Mass Spectral Interpretation

A. Primary Aliphatic Alcohols

1. General formula
   ROH

2. Molecular ion:
   The intensity of the molecular ion in both straight-chain and branched alcohols decreases with increasing molecular weight. Beyond C_5, in the case of branched primary alcohols, and C_6, in the case of straight-chain primary alcohols, the molecular ion is usually insignificant.
   The peak representing the loss of water from the molecular ion can easily be mistaken for the molecular ion. The spectrum is similar to an olefin below the $[M - H_2O]^+$ peak except that the peaks at m/z 31, 45, and 59 indicate an oxygen-containing compound.

3. Fragmentation
   A primary alcohol is indicated when the m/z 31 peak is intense and will be the base peak for C_1–C_4 straight-chain primary alcohols. C_4 and higher straight-chain primary alcohols lose 18, 33, and 46 Daltons from the molecular ion. Branched aliphatic alco-

hols do not appear to lose 46 Daltons from the molecular ion. Branching at the end of the chain, especially when an *iso*propyl or *tert*-butyl group is present, results in intense peaks at masses 43 and 57. In addition, a peak is observed corresponding to a loss of 15 Daltons from the molecular ion.

4. Characteristic fragment ions
   m/z 19 and 31
   m/z 41, 55, etc., similar to 1-olefins.

5. Characteristic losses from the molecular ion
   M − 18, M − 33, M − 46

B. Secondary and Tertiary Alcohols

1. General formula
   R_2CHOH and R_3COH

2. Molecular ion
   The molecular ion is slightly more intense in the mass spectra of secondary alcohols than in tertiary alcohols, but even in secondary alcohols, the molecular ion intensity is very small.

3. Fragmentation
   Many low-molecular weight ($<C_8$) secondary and tertiary alcohols exhibit no M − 18 peaks. C_8 and higher secondary alcohols exhibit M − 18 peaks. The M − 46 peak is usually missing in the mass spectra of secondary and tertiary alcohols. In secondary and tertiary alcohols, the loss of the largest alkyl group results in intense fragment ions.

4. Characteristic fragment ions

   M − 18 $>C_8$

   M − 33

   no M − 46

Mass 45 for secondary alcohols with a methyl on the α-carbon,

$$\begin{array}{c} R \\ | \\ ----|----\;_{45} \\ | \\ CH_3CHOH \end{array}$$

Mass 59 for tertiary alcohols with two methyl groups on the α-carbon,

$$\begin{array}{c} R \\ \text{----}\overset{|}{\underset{|}{}}\text{----}^{59} \\ (CH_3)_2\,COH \end{array}$$

C. Cyclic Alcohols

   1. General formula

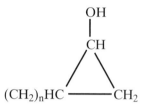

   2. Molecular ion
       The intensity of the molecular ion is generally less than 2.5%.

   3. Fragmentation
       The intensity of the m/z 31 ion is sufficient to suggest the presence
       of oxygen. Masses 44 and 57 are usually present, and an M − 18
       peak is also detectable. Mass 44 usually suggests an aldehyde
       unbranched on the α-carbon, but this ion is also prominent in
       the mass spectra of cyclobutanol, cyclopentanol, cyclohexanol,
       and so forth. Mass 57 (C_3H_5O) is also fairly intense for C_5 and
       larger cyclic alcohols. If an aldehyde is present, M − 1, M − 18,
       and M − 28 peaks are observed.

   4. Characteristic fragment ions

       M − 18

       Masses 44 and 57 are fairly intense.

       No M − 1 or M − 28.

D. Mass Spectra of TMS Derivatives of Aliphatic Alcohols
    The mass spectrum of the trimethylsilyl derivative is used to deter-
    mine the molecular weight of the unknown alcohol, even though
    the molecular ion may not be observed. If two high-mass peaks
    are observed and are 15 mass units apart, then the highest mass
    (excluding isotopes) is the molecular ion of the TMS derivative.
    However, if only one high-mass peak is observed, add 15 mass units
    to deduce the molecular ion. The molecular weight of the alcohol
    is determined by subtracting 72 (C_3H_8Si) from the molecular ion

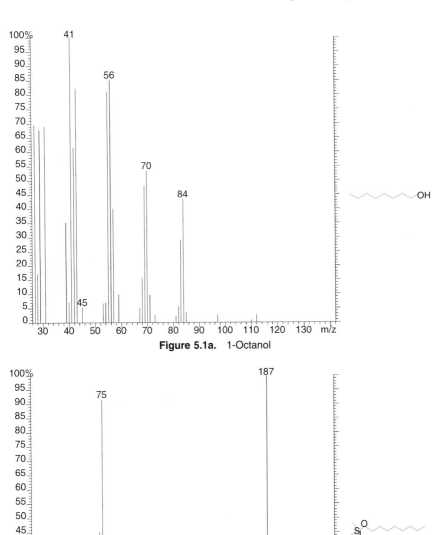

Figure 5.1a. 1-Octanol

Figure 5.1b. TMS Derivative 1-Octanol

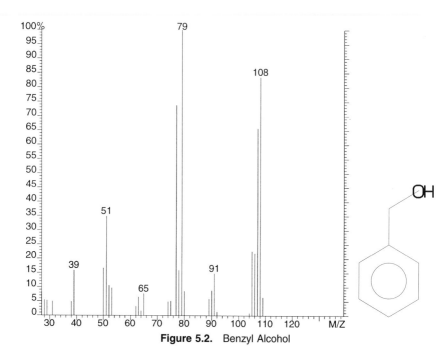

Figure 5.2. Benzyl Alcohol

of the TMS derivative. Ions at *m/z* 73, 89, and 103 are also usually present in the mass spectra of the TMS derivatives of aliphatic alcohols.

E. Sample Mass Spectra

1. In the mass spectrum of 1-octanol (Figure 5.1a) peaks at *m/z* 31 and 45 show that the compound contains oxygen. The presence of an intense *m/z* 31 peak further suggests that it is a primary aliphatic alcohol, ether, or possibly a ketone. By adding 18 Daltons to the highest-mass ion observed, the deduced molecular weight would be 130. Now check to see if M− 33 and M − 46 are present (at *m/z* 97 and 84). This mass spectrum suggests a primary aliphatic alcohol with a molecular weight of 130, which is 1-octanol (caprylic alcohol), $C_8H_{18}O$.

2. For the TMS derivative of 1-octanol (Figure 5.1b), note the large *m/z* 187 ion and the small ion 15 Daltons higher in mass. The molecular ion of the TMS derivative is at *m/z* 202. Subtract 72 from 202 to obtain the molecular weight of the alcohol.

3. The ions at *m/z* 77, 65, 51, and 39 in Figure 5.2 suggest a phenyl group. The ion at *m/z* 91 suggests a benzyl group, and themolecular ion 17 Daltons higher in mass suggests benzyl alcohol.

IV. Aminoalcohols (See Chapter 8, Amines)

C h a p t e r   6

Aldehydes

I. GC Separation of Underivatized Aldehydes

A. Capillary Columns

1. Acetic acid, *iso*butyraldehyde, methylethyl ketone, *iso*butyl alcohol, *n*-propyl acetate, and *iso*butyric acid
30 m Poraplot Q column, 100–200° at 10°/min.

2. Acetaldehyde, acetone, tetrahydrofuran (THF), ethyl acetate, *iso*propyl alcohol, ethyl alcohol, 4-methyl-1,3-dioxolane, *n*-propyl acetate, methyl *iso*butyl ketone, *n*-propyl alcohol, toluene, *n*-butyl alcohol, 2-ethoxyethanol, and cyclohexane
30 m DB-WAX column, 75° (16 min)–150° at 6°/min.
Although the DB-FFAB column is similar to the DB-WAX column, it should not be used to separate aldehydes because it may remove them from the chromatogram.

3. Acetaldehyde, acetone, *iso*propyl alcohol, ethyl acetate, methyl *iso*butyl ketone, toluene, butyl acetate, *iso*butyl alcohol, and acetic acid
30 m FFAP-DB column, 50–200° at 6°/min.

4. a. Aromatic aldehydes
Benzyl alcohol, 1-octanol, benzaldehyde, octanoic acid, benzophenone, benzoic acid, and benzhydrol
30 m DB-WAX column, 60° (1 min)–230° at 10°/min.

    b. Tolualdehydes
    Ortho- and *meta*-isomers do not separate very well. *Para*-isomers elute last.
    50 m DB-Wax column, 60–80° at 6°/min.

B. Packed Columns

  1. a. Acetaldehyde, furan, acrylic aldehyde, propionaldehyde, *iso*-butyraldehyde, *n*-butyraldehyde, and 2-butenal
     2 m Porapak N column, 50–180° at 6°/min.

    b. Formaldehyde, water, and methanol
     2 m Porapak N column at 125° (not for trace analyses).

  2. Acetaldehyde, methanol, acetone, ethanol, *iso*propyl alcohol, and *n*-propyl alcohol
    2 m CW 20M column on Carbopack B at 75°.

  3. Acetaldehyde, methanol, ethanol (major), ethyl acetate, *n*-propyl alcohol, *iso*butyl alcohol, acetic acid, amyl alcohol, and *iso*-amyl alcohol
    2 m CW 20M column on Carbopack B, 70–170° at 5°/min.

  4. Acetone, acrolein, 2,3-dihydrofuran, butyraldehyde, *iso*propyl alcohol, tetrahydrofuran, 1,3-dioxolane, 2-methyltetrahydrofuran, benzene, and 3-methyltetrahydrofuran
    2 m 3% SP-1500 column on Carbopack B, 60–200° at 6°/min.

II. Derivatization of Formaldehyde

A. Formaldehyde is derivatized for trace analyses. React 2-hydroxymethylpiperidine with formaldehyde to form 3,4 tetramethyleneoxazole ($C_7H_{13}NO$).
   Selected ion monitoring of mass 127 is used to determine the concentration of formaldehyde.

III. Mass Spectra of Aldehydes

A. Aliphatic Aldehydes

  1. General formula
    RCHO

  2. Molecular ion
    Both straight-chain and branched aliphatic aldehydes show molecular ion peaks up to a minimum of C_{14} aldehydes.

3. Fragmentation

Above C_4, aliphatic aldehydes undergo the McLafferty rearrangement, resulting in an observed m/z 44 ion, provided the α-carbon is not substituted. Substitution on the α-carbon results in a higher m/z peak. (See the proceeding text.)

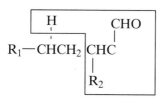

Note: Subtract 43 from the rearrangement ion to determine R_2.

When $R_2 = H$, observe m/z 44.
When $R_2 = CH_3$, observe m/z 58.
When $R_2 = C_2H_5$, observe m/z 72, etc.
Small peaks at masses 31, 45, and 59 indicate the presence of oxygen in the compound. Also, aldehydes lose 28 and 44 Daltons from their molecular ions.

4. Characteristic fragment ions

The mass spectra of aliphatic aldehydes show m/z 29 (CHO) for C_1–C_3 aldehydes and m/z 44 for C_4 and longer chain aldehydes. Characteristic losses from the molecular ion:

M − 1 (H)

M − 18 (H_2O)

M − 28 (CO)

M − 44 (CH_3CHO)

Aldehydes are distinguished from alcohols by the loss of 28 and 44 Daltons from the molecular ion. The M − 44 ion results from the McLafferty rearrangement with the charge remaining on the olefinic portion.

B. Aromatic Aldehydes

1. General formula
ArCHO

2. Molecular ion
Aromatic aldehydes give a very intense molecular ion.

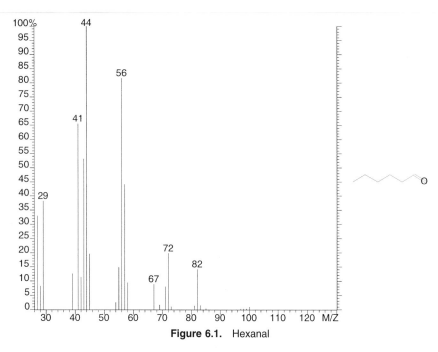

Figure 6.1. Hexanal

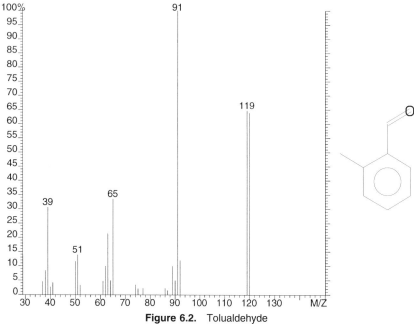

Figure 6.2. Tolualdehyde

3. Fragmentation

The M − 1 peak due to the loss of the aldehyde hydrogen by α-cleavage is usually abundant. The loss of 29 Daltons is characteristic of aromatic aldehydes. Peaks at m/z 39, 50, 51, 63, and 65 and the abundance of the molecular ion show that the compound is aromatic. Accurate mass measurement data indicate the presence of an oxygen atom.

4. Characteristic losses from the molecular ion

    M − 1 (H)

    M − 29 (CHO)

C. Sample Mass Spectra

1. An intense ion at m/z 44 in the mass spectrum of hexanal suggests an aliphatic aldehyde. M − 18, M − 28, and M − 44 (at m/z 56, 72, and 82, respectively) suggest an aliphatic aldehyde unbranched at the α-carbon (see Part III, Ions for Determining Unknown Structures). The molecular ion at m/z 100 confirms that this is the spectrum of hexanal (see Figure 6.1).

2. The mass spectrum of 2-methylbenzaldehyde suggests an aromatic compound because of the intensity of the molecular ion and peaks at m/z 39, 51, and 65 (see Figure 6.2). The loss of hydrogen atoms and loss of 29 Daltons from the molecular ion indicate that this is an aromatic aldehyde. Looking up m/z 91 in Part III suggests the following structure:

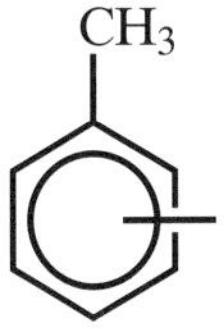

From the molecular ion (m/z 120), the structure for the aromatic aldehyde is

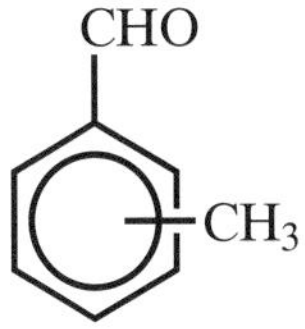

C h a p t e r   7

Amides

I. GC Separation of Underivatized Amides

A. Capillary Columns

   1. General conditions for separation of amides
   30 m FFAP-DB column, 80–220° at 12°/min.

   2. Hexamethylphosphoramide, pentamethylphosphoramide, tetra-methylphosphoramide, trimethylphosphoramide, and
   $[(CH_3)_2N]_2 \, P(O)NHCHO$
   30 m DB-WAX column, 60–220° at 10°/min.

   3. N,N-dimethylacetamide (DMAC) impurities:
   N,N-dimethylacetonitrile, DMF, DMAC, N-methylacetamide, and acetamide
   60 m DB-WAX column, 60–200° at 7°/min.

B. Packed Columns

   1. Acrylamide impurities (dissolve the sample in a minimum amount of methanol) also separate formamide, acetamide, dimethylformamide, N-methylacetamide, propionamide, N,N-dimethylacetamide
   2 m Tenax-GC column, 100–170° at 3°/min.

II. Derivatization of Amides

A. Primary Amides

Derivatized primary amides are more volatile. Typically TMS or *N*-dimethylaminomethylene derivatives are prepared.

1. TMS derivatives of amides
   Add 250 μl of Tri-Sil/BSA (Formula P) reagent to less than 1 mg of the sample. Heat at 60° for 15–20 min.

$$RC(O)NH_2 \longrightarrow RC(O)NHSi(CH_3)_3$$

2. *N*-dimethylaminomethylene derivatives of primary amides
   Add 250 μl of Methyl-8 reagent to less than 1 mg of sample. Heat at 60° for 20–30 min.

$$RC(O)NH_2 \xrightarrow{\text{Methyl-8}} R \overset{99}{\underset{|}{|}} C(O)N = CHN(CH_3)_2$$

This derivative also works well with diamines or amino amides (e.g., 6-aminocaproamide).

$$H_2N(CH_2)_5C(O)-NH_2$$

$$\xrightarrow{\text{Methyl-8}} (CH_3)_2NCH{=}N(CH_2)_5C(O)N{=}CHN(CH_3)_2$$

B. Aromatic Amides (ArC(O)NHR or RC(O)NHAr, where R can represent H, an alkyl group, or an aromatic group.)

1. Derivatization of aromatic amides
   Except for simple aromatic amides such as benzamide and acetanilide, derivatization is recommended.[*] The most common derivatives used in this laboratory are TMS, acetate, and *N*-dimethylaminomethylene (for primary amides).

   a. Preparation of TMS derivatives of primary and secondary aromatic amides
      Add 250 μl of TRI-SIL/BSA (Formula D) reagent to less than 1 mg of sample. Heat at 60° for 15–30 min.

   b. Preparation of acetate derivatives of primary and secondary aromatic amides

---

*Even when they are underivatized, benzamide and benzanilide have very similar mass spectra.

Add 150 μl of acetic anhydride and 100 μl of pyridine to less than 1 mg of sample. Heat at 60° for 30 min. Evaporate to dryness with clean, dry nitrogen. Dissolve residue in 25 μl of DMF or other suitable solvent.

  c. *N*-dimethylaminomethylene derivative of primary aromatic amides
Add 250 μl of Methyl-8 reagent to less than 1 mg of sample dissolved in DMF if necessary. Heat at 60° for 20–30 min.

III. GC Separation of Derivatized Amides (TMS or Methyl-8)

A. 30 m DB-5 column, 60–275° at 8°/min.

B. Aromatic Amides
30 m DB-5 column, 150–300° at 10°/min.

IV. Mass Spectra of Amides

A. Primary Amides

1. General formula
RC(O)NH$_2$

2. Molecular ion
The mass spectra of underivatized amides generally show molecular ions.

3. Fragmentation
For straight-chain amides (>C$_3$), having a γ-hydrogen, the base peak is *m/z* 59 (C$_2$H$_5$NO), which occurs by a McLafferty rearrangement.

$$\text{RCHCH}_2\text{CH}_2\text{C-NH}_2 \longrightarrow [\text{C}_2\text{H}_5\text{NO}]^+ + \text{RCH}=\text{CH}_2$$
m/z 59

Ions are also observed at *m/z* 44, 58, 72, etc., with *m/z* 59 and 72 being the most intense.

72   58   44
R｜CH$_2$｜CH$_2$｜C(O)NH$_2$

In summary, if the unknown mass spectrum has an intense peak at m/z 59 and an abundant m/z 72 with an odd molecular ion, this suggests a primary amide.

B. Secondary Amides

1. General formula
   $R_1C(O)NHR_2$

2. Molecular ion
   As would be expected, the molecular ion decreases in intensity as R_1 or R_2 increases in length.

3. Fragmentation
   The mass spectra of secondary amides have an intense rearrangement ion at m/z 30. The fragmentation of the R_2 chain can occur to yield $R_1C(O)NH_2^{\cdot+}$ or $R_1C(O)NH_3^+$.
   Secondary amides also undergo the McLafferty rearrangement:

$$\left[R_1C \overset{\displaystyle O}{\diagup} NH\text{-}R_2 \right]^{+\cdot} \longrightarrow \left[CH_3C(O)NHR_2 \right]^{+\cdot}$$

   Subtract 58 Daltons from this rearrangement ion to find R_2.

C. Tertiary Amide

1. General formula
   $R_1C(O)NR_2R_3$

2. Molecular ion
   A molecular ion is observed when R_1, R_2, and R_3 are less than or equal to C_4.

3. Fragmentation
   Providing the R_1 group has a γ-hydrogen, $[CH_3C(O)NR_2R_3]^{\cdot+}$ is a common fragment. Subtract 57 Daltons from this rearrangement ion to find $R_2 + R_3$.

D. Mass Spectra of Aromatic Amides
   In simple aromatic amides, fragmentation occurs on both sides of the carbonyl group. If a hydrogen is available in *N*-substituted aromatic amides, it tends to migrate and form an aromatic amine

and the loss of a ketene. Some simple aromatic amides include: benzamide, dibenzamide, *N*-phenyldibenzamide, nicotinamide, *N,N*-diethylnicotinamide, acetanilide, and benzanilide.

Benzamide (MW 121)

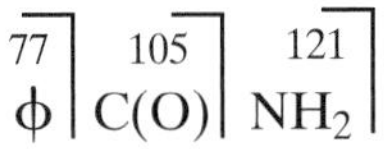

$$\overline{77\mid}\ \overline{105\mid}\ \overline{121\mid}$$
$$\phi\mid C(O)\mid NH_2\mid$$

m/z 105, 77, 121

Dibenzamide (MW 225)

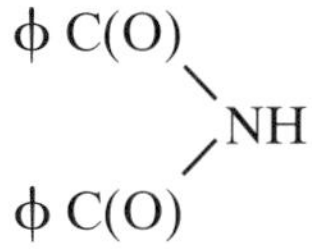

$$\phi\ C(O)$$
$$\searrow NH$$
$$\phi\ C(O) \nearrow$$

m/z 105, 77, 225

N-Phenyldibenzamide (MW 301)

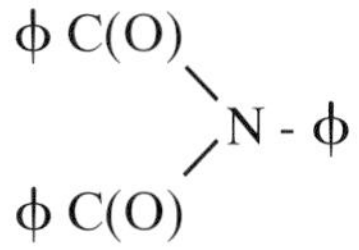

$$\phi\ C(O)$$
$$\searrow N - \phi$$
$$\phi\ C(O) \nearrow$$

m/z 105, 77, 197

Nicotinamide (MW 122)

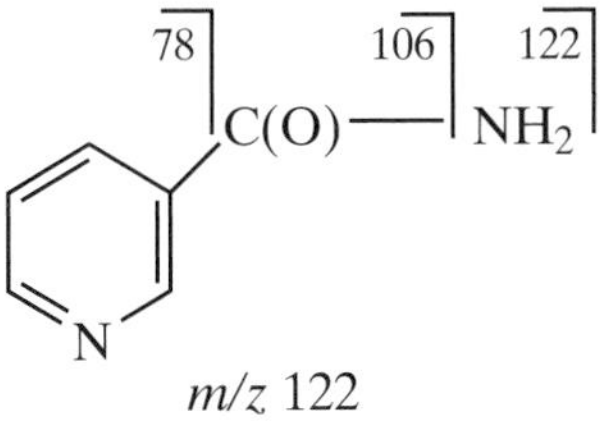

m/z 122

N,N-Diethylnicotinamide (MW 178)

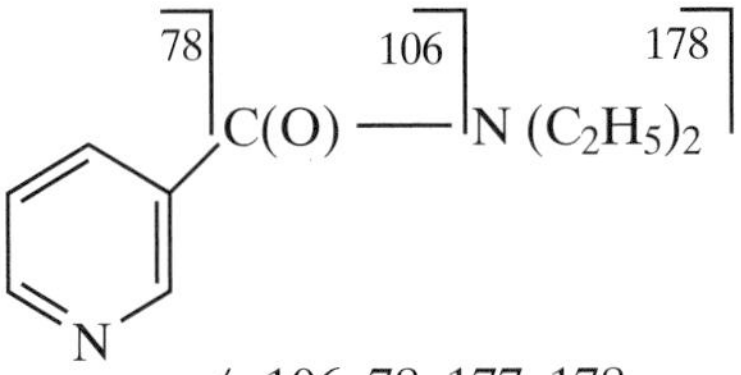

m/z 106, 78, 177, 178

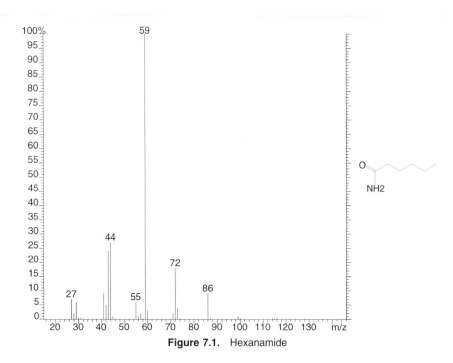

Figure 7.1. Hexanamide

Acetanilide (MW 135)

$CH_3C(O)NHPh$
m/z 93, 66, 135

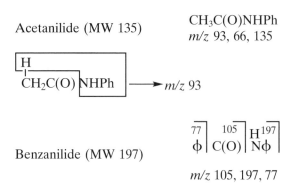

Benzanilide (MW 197)

$$\overline{77}\Big|\overset{105}{}\Big|\overset{197}{H}$$
$$\phi\Big|C(O)\Big|N\phi$$

m/z 105, 197, 77

E. Sample Mass Spectrum

Notice the ions at *m/z* 59, 44, and 72 in the mass spectrum of hexanamide (Figure 7.1). Looking these up in the tables in Part III suggests an aliphatic primary amide. Looking at the mass spectrum very closely, the highest-mass ion occurs at *m/z* 115. These data suggest a hexanamide.

V. Mass Spectra of Derivatized Amides

Benzamide-TMS (MW 193)   *m/z* 73, 178, 135, 193, 192

Acetanilide-acetate (MW 177)   *m/z* 93, 135, 43, 77

C h a p t e r   8

Amines

I. GC Separations of Underivatized Amines

Although capillary columns are generally preferred, there are many examples where separation is better using packed columns, especially for low-boiling amines.

A. Low-Boiling Aliphatic Amines

   1. Amines from C_1 (methylamine) to C_6 (cyclohexylamine)
      2 m 4% Carbowax 20M column + 0.8% KOH on 60–80 mesh Carbopack B, 75–150° at 6°/min.

   2. *Iso*propylamine, *n*-propylamine, di*iso*propylamine, di-*n*-propylamine
      2 m Chromosorb 103 column, 50–150° at 8°/min.

   3. Methylamine, ethylenimine, dimethylamine, trimethylamine
      2 m Chromosorb 103 column, 60–180° at 6°/min.

   4. 1-Aminooctane, 2-aminooctane
      2 m Chromosorb 103 column at 135°.

B. Higher-Boiling Aliphatic Amines and Diamines

   1. Diaminoethane, diaminopropane, diaminobutane, diaminopentane, diaminohexane, diaminooctane

25 m CPWAX column for amines and diamines (Chrompack cat. no. 7424), 75-200° at 6°/min.

2. Diaminoethane, diaminopropane, 1-amino-2-propanol, diaminobutane, diaminopentane, *n*-decylamine
   25 m CPWAX column for amines and diamines at 135°.

C. Aromatic Amines and Diamines

1. Aniline, 2,3,4-picolines
   2 m Carbowax 20M column or Carbopack B column, 75–150° at 3°/min.

2. Dimethylanilines and trimethylanilines
   25 m DB-1701 column, 60–270° at 5°/min.

3. Toluidine, nitrotoluene isomers, diaminotoluene, and dinitrotoluene isomers
   30 m DB-17 column, 100–250° at 8°/min.

4. Phenylenediamines

   a. Lower-boiling impurities (sample dissolved in acetonitrile)
      Aniline (MW 93), quinoxaline (MW 130), dimethylquinoxaline (MW 158) from the phenylenediamines (MW 108)
      25 m CP-WAX column for amines (Chrompack) at 200°.

   b. Higher-boiling impurities
      Some impurities may be found under these GC conditions: quinoxaline, phenazine, tetrahydrophenazine, nitroanilines, hydroxyanilines, chloronitrobenzenes, hydroquinone, diaminophenazine, aminohydroxyphenazine
      30 m DB-17 column, 100–275° at 6°/min.

5. Di*iso*propylamine, di*iso*butylamine, dibutylamine, pyridine, dicyclohexylamine, aniline, 2,6-dimethylaniline
   25 m CP Wax 51 column, 70-210° at 5°/min.

II. Derivatization of Amines and Diamines

MTBSTFA is the recommended reagent for silylating the amine functionality because it forms a more stable derivative than MSTFA, BSTFA, or BSA. Because amines can be difficult to silylate, the solvent used is important.

A. TBDMS and TMS Derivatives
   Add 0.1 mg of the sample in 50 μl of acetonitrile (or THF) to 50

μl reagent. Let the solution stand at room temperature for 10–20 min.

> Reagents: MTBSTFA (recommended)
> MSTFA (recommended for amine hydro-chlorides)
> BSTFA
> BSA

B. Preparing Methyl-8 Derivatives
Add less than 0.1 mg of sample to 50 μl of acetonitrile and then add 50 μl of methyl-8 reagent [$(CH_3)_2NCH(OCH_3)_2$]. Heat at 60° for 30 min or at 100° for 20 min.

III. GC Separation of Derivatized Amines

A. Diamines

1. TBDMS or TMS derivatives of diamines
30 m DB-225 column, 75–225° at 8°/min.

2. Methyl-8 derivatives of diaminohexanes and diaminooctanes
30 m DB-5 column, 80–270° at 8°/min.

IV. Mass Spectral Interpretation of Amines

A. Underivatized
Organic compounds with an odd number of nitrogen atoms will have an odd-mass molecular ion and prominent fragment ions at even masses. For amines, the most important fragmentation is cleavage of the bond that is β to the nitrogen atom with the charge remaining on the nitrogen-containing fragment.

Underivatized diamines are difficult to identify by their mass spectra alone because of the low abundance of the molecular ion (<3%). However, M − 17 is a common fragment ion. β-cleavage is prominent in diamines.

1. Primary amines
Primary amines show characteristic peaks at masses 18 [NH_4]$^+$ and 30 [$CH_2{=}NH_2$]$^+$. If the α-carbon is alkyl substituted, then intense ions are observed at m/z 44, 58, or 72, and so on. If the unknown amine reacts with acetone or Methyl-8 then it is a primary amine.

Example 8.1

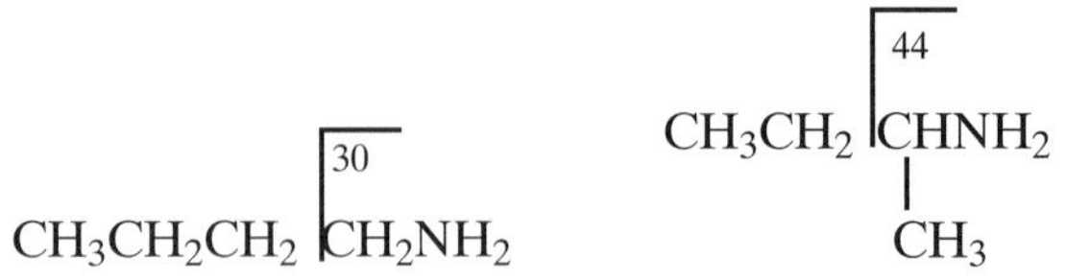

$$CH_3CH_2CH_2 \overline{\left\lceil {}^{30} \right.} CH_2NH_2$$

$$CH_3CH_2 \overline{\left\lceil {}^{44} \right.} \underset{\underset{CH_3}{|}}{CHNH_2}$$

2. Secondary amines

With secondary amines, cleavage of the bond β to the nitrogen atom occurs preferentially at the shortest hydrocarbon chain. If the shortest hydrocarbon chain has three or more carbon atoms, α, β-cleavage occurs with a hydrogen rearrangement.

Example 8.2

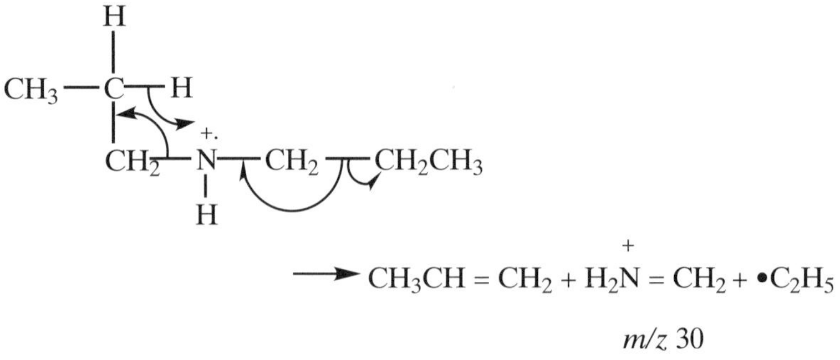

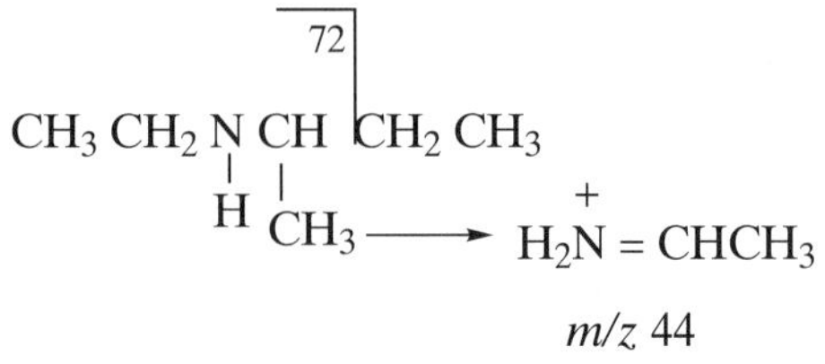

$$\longrightarrow CH_3CH = CH_2 + H_2\overset{+}{N} = CH_2 + \bullet C_2H_5$$

$$m/z\ 30$$

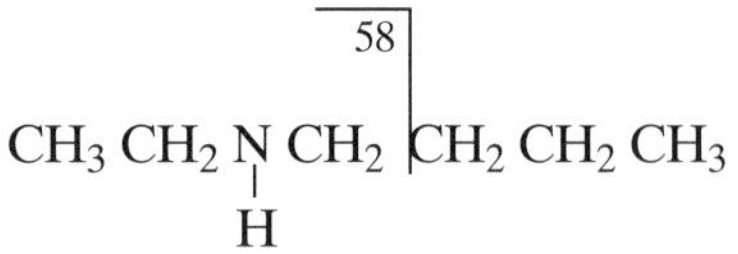

$$CH_3\ CH_2\ \underset{\underset{H}{|}}{N}\ CH_2 \overline{\left\lceil {}^{58} \right.} CH_2\ CH_2\ CH_3$$

$$CH_3\ CH_2\ \underset{\underset{H}{|}}{N}\ \underset{\underset{CH_3}{|}}{CH} \overline{\left\lceil {}^{72} \right.} CH_2\ CH_3 \longrightarrow H_2\overset{+}{N} = CHCH_3$$

$$m/z\ 44$$

3. Tertiary amines

Tertiary amines undergo β-cleavage, preferentially in the longest hydrocarbon chain. If more than three carbon atoms are present, then α, β-cleavage along with hydrogen rearrangement occurs.

$$R_1CH_2CH_2NCH_2R_3 \overset{+\cdot}{\underset{R_2}{\rceil}} \longrightarrow R_1CH=CH_2 + R_2\overset{+}{N}H=CH_2 + \bullet R_3$$

Summary: If the molecular weight is odd, then the compound contains an odd number of nitrogens. Fragment ions observed at even-mass numbers suggests the presence of nitrogen. The loss of ammonia is fairly common in nitrogen compounds and may not indicate exclusively that an amine is present. Chemical derivatization will easily determine if the unknown is a tertiary amine.

4. Mass spectra of cyclic amines
   Molecular ions as well as $[M - H]^+$ ions are observed for cyclic amines.

Ethylenimines		$[M - H]^+$ $[M - CH_3]^+$
Pyrrolidines		$[M - H]^+$ $[M - 28]^+$
Piperidines		$[M - H]^+$ $[M - 29]^+$ (*m/z* 84 is present in alkylpiperidines)
Hexamethyleneimines		$[M - H]^+$ $[M - 29]^+$ (*m/z* 112 is present in alkylhexamethyleneimine)

Mass 28 in cyclic amines is CH_2N. Mass 30 is fairly intense in nonsubstituted cyclic amines.

5. Mass spectra of cycloakylamines
   Molecular ions are easily detected for most cycloalkylamines and have a characteristic *m/z* of 30. For methylcyclopentylyamine, *m/z* 30 is the base peak in the spectrum, whereas for cyclohexylamine a rearrangement fragment ion is the base peak.

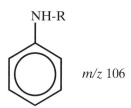

$$CH_3NH \text{(cyclopentyl)}^{+\bullet} \longrightarrow [CH_4N]^+ + \bullet C_5H_9$$

$$\text{(cyclohexyl)}NH_2^{+\bullet} \longrightarrow \text{(fragment)}NH_2^+ \bullet \longrightarrow CH_2CHCHNH_2^+ + \bullet C_3H_7$$

6. Mass spectra of aromatic amines
   Aromatic amines show intense molecular ions. When alkyl side chains are present, the molecular ions decrease with increasing alkyl chain length, but the molecular ions are still fairly intense. Aromatic amines (including the naphthylamines) lose 1, 27, and 28 Daltons from their molecular ions, but these losses also decrease in intensity as the alkyl side chain increases in size. From the mass spectrum alone it is difficult to determine whether the alkyl group is on the ring or on the nitrogen. An *m/z* 106 represents an intense ion when one alkyl group is on the nitrogen:

NH-R

(benzene ring) *m/z* 106

7. Mass spectral fragmentation

Amine	Characteristic Fragments	Rearrangement Ions	Characteristic Losses from the Molecular Ion
RNH_2	*m/z* 30	*m/z* 18 (NH_4)	M − NH_3 (especially diamines)
R∖ ⁄NH ⁄R	β-Cleavage longest chain	*m/z* 30 (NH_2CH_2)	
R∖ R—N ⁄R	β-Cleavage longest chain	$(NHRCH_2)^+$	

B. Derivatized

Preparing derivatives of amines can make the identifications much easier.

Functional Group	Derivative	Increase in MW
$-NH_2$	$-NH-Si(CH_3)_3$ (TMS)	72
$-(CH_2)_3NH_2$	$-N[Si(CH_3)_3]_2$ (TMS)	144
$-NH_2$	$NH-Si(CH_3)_2C(CH_3)_3$ (TBDMS)	114
$-NH_2N$	$HCOCF_3$ (Trifluoroacetyl)	96
$-NH_2$	$N=CHN(CH_3)_2$ (Methyl-8)	55

If at least three CH_2 groups are present between the amino group and another functional group, it is possible to add two TMS groups to the amine functional group. The presence of an intense ion at *m/z* 174 confirms the addition of two TMS groups on the same nitrogen. By adding mass 57 to the last intense ion, the molecular weight of the compound is determined including the TBDMS derivative. The Methyl-8 derivative is excellent for use in analyzing diaminohexanes, diaminooctanes, and so forth. Only one Methyl-8 derivative adds per nitrogen so that one does not obtain multiple derivatives as with TMS. The molecular ions are relatively intense.

C. Sample Mass Spectrum

The most prominent ion in the mass spectrum of 1-octanamine is *m/z* 30 (Figure 8.1). From Part III, a very intense *m/z* 30 suggests

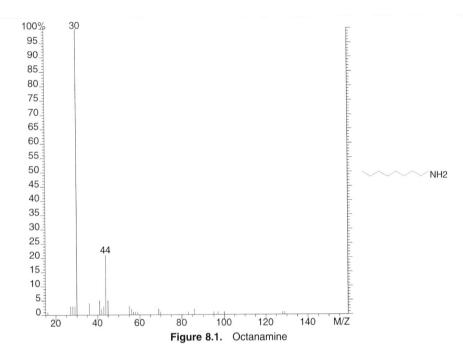

Figure 8.1. Octanamine

a primary amine, secondary amine, or a nonsubstituted cyclic amine.
A very small molecular ion occurs at *m/z* 129. If the unknown reacts
with Methyl-8 reagent, it is a primary amine (not a secondary amine)
or a cyclic amine.

$$CH_3(CH_2)_7NH_2 \xrightarrow{\text{Methyl-8}} CH_3(CH_2)_7N = CHN(CH_3)_2$$

MW 129   MW 184

V. Amino Alcohols (Aliphatic)

A. GC Separation of Underivatized Amino Alcohols

1. Monoethanolamine (MEA), diethanolamine (DEA), trietha-
nolamine, and impurities
30 m, 1.0 μm Rtx-35 (Restek) column, 40 (2 min)–250° at
6°/min (hold 15 min).

2. 1-Amino-2-propanol, 3-amino-1-propanol, 2-amino-2-methyl-1-
propanol, and similar compounds

25 m CPWAX-51 column or CPWAX column for amines, 50-210° at 5°/min.

B. GC Separation of Derivatized Amino Alcohols

1. 30 m DB-1701 column, 45 (10 min)-250° at 10°/min.

C. Mass Spectra of Amino Alcohols
In the mass spectra of amino alcohols, the m/z 30 peak is intense while the m/z 31 peak is of relatively low abundance. The amino group dominates the fragmentation, making it difficult to recognize the alcohol group. If a m/z 30 peak is found in the mass spectrum of an unknown, it does not mean that no alcohol group is present. The loss of 31 Daltons from the molecular ion suggests the presence of a terminal alcohol group.

To identify the presence of an amino alcohol use the following procedure. Prepare a TMS derivative of the unknown using MSTFA reagent and obtain a mass spectrum of the resulting TMS derivative. Prepare a second TMS derivative of the unknown using TRI-SIL Z reagent. When the molecular weight of the unknown increases by 144 mass units using MSTFA reagent and the TMS derivative using TRI-SIL Z reagent only increases the molecular weight by 72 mass units, this suggests the presence of both an amino group and a hydroxyl group in the unknown. TRI-SIL Z reagent silylates alcohols and carboxy hydroxyl groups, but not amino groups.

A method to determine the number of amino groups present in the molecule requires the formation of a TMS derivative with MSTFA, which silylates both hydroxyl and amino groups.[1] First, obtain a mass spectrum of the unknown using GC/MS. Next, add MBTFA reagent to the previously prepared TMS reaction mixture and let stand approximately 30 minutes. Obtain a mass spectrum of the resulting reaction product. A trifluoroacetyl group will replace each TMS group on primary and secondary amino groups, because the amino-TMS group is less stable. From the mass differences obtained before and after the reaction with MBTFA (24 for each amino group), the number of amino groups present can be determined.

MBTFA $CF_3CON(CH_3)COCF_3$
     N-methyl-bis (trifluoroacetamide)

[1]Sullivan J., and Schewe, L. *J. Chromat. Sci.*, *15*, 196–197, 1977.

MSTFA $CF_3CON(CH_3)$-TMS
N-methyl-*n*-trimethylsilyl trifluoroacetamide

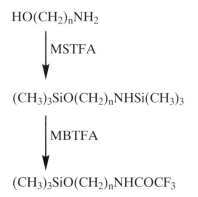

$HO(CH_2)_nNH_2$

MSTFA

$(CH_3)_3SiO(CH_2)_nNHSi(CH_3)_3$

MBTFA

$(CH_3)_3SiO(CH_2)_nNHCOCF_3$

VI. Aminophenols

A. GC Conditions

1. Aminophenols (underivatized)
   o-aminophenol, *p*-aminophenol, and acetanilide
   30 m DB-1701 column, 45–250° at 10°/min.

2. Aminophenols (as acetates; see Chapter 32, Section IC, for acetate derivatization procedure)
   Acetanilide (C_8H_9NO), *o*-aminophenol, and *p*-aminophenol ($C_{10}H_{11}O_3N$)
   30 m DB-1 column, 100–250° at 10°/min.

B. Mass Spectra of Underivatized Aminophenols
   Molecular ions of unsubstituted aminophenols are intense. The M − 28 and M − 29 ions are observed. Prominent ions are *m/z* 80 and 109.

C. Mass Spectra of Aminophenols as Acetates
   The molecular ions of the unsubstituted phenols are present, but are smaller than the underivatized aminophenols. The M − 42 is characteristic of acetates. Prominent ions are *m/z* 109, 151, and 193.

C h a p t e r   9

Amino Acids

I. GC Separation

A. GC Separation of TBDMS Derivatized Amino Acids
   30 m DB-1 (Methylsilicone) column, 150–250° at 3°/min, then 250–275° at 10°/min. Run time is 1 hour.
   Retention times and suggested ions for selected ion monitoring (SIM) of TBDMS derivatives of AAs are given in Table 9.1 and the GC separation is shown in Figure 9.1. Amino acids should be derivatized prior to separation. The TBDMS derivatives are preferred and stable for at least 1 week.

B. GC Separation of PTH-Amino Acids

   1. The only PTH-AAs that can be analyzed easily by GC/MS without derivatization are alanine, glycine, valine, proline, leucine, isoleucine, methionine, and phenylalanine
      30 m DB-1 column, 75–275° at 8°/min. Dissolve the PTH-AA in the minimum amount of ethyl acetate.

   2. Order of elution and molecular weight (MW)
      Alanine (206), glycine (142), valine (234), proline (232), leucine (248), isoleucine (248), methionine (266), and phenylalanine (282).

87

Table 9.1.   Retention times and accurate masses for TBDMS derivatized amino acids (separation conditions Chapter 9.1.A)

Approximate retention time (min)	Amino acid	For selected ion monitoring
5:55	Alanine	158.1356, 232.1553, 260.1502
6:17	Glycine	218.1396, 246.1346
8:28	Valine	186.1678, 228.1815, 260.1866
9:24	Leucine	200.1834, 302.1972, 274.2022
10:12	Isoleucine	200.1834, 302.1972, 274.2022
10:51	Proline	184.1522, 286.1659, 258.1709
16:00	Methionine	292.1587, 320.1536
17:10	Serine	390.2316
18:06	Threonine	376.2523, 404.2473
19:29	Phenylalanine	234.1678, 336.1815
21:40	Aspartic acid	418.2265
22:40	Hydroxyproline	314.2335, 416.2473
22:58	Cysteine	406.2088
25:00	Glutamic acid	432.2422
25:35	Asparagine	417.2425
28:11	Lysine	300.1815, 431.2945
28:55	Glutamine (peak 1)	431.2581
30:52	Arginine	442.2741
33:03	Histidine	338.2448, 440.2585
34:17	Glutamine (peak 2)	413.2476
34:34	Tyrosine	466.2629
38:51	Tryptophan	489.2789
45:57	Cystine	348.1849

C. GC Conditions for TBDMS Derivatized PTH-Amino Acids*
15 m DB-5 column (or equivalent), 80–300° at 15°/min.

D. GC Conditions for N-PFP Isopropyl Esters of D and L Amino Acids
25 to 50 m CHIRASIL-L-VAL column (Chrompack cat. no. 7495),

*If the reaction time for the TBDMS derivatives is not long enough, a mixture of mono- and di-TBDMS derivatives is observed, resulting in more than one GC peak and thus reduced sensitivity.

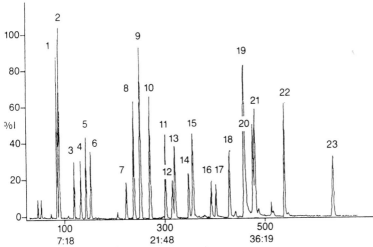

Figure 9.1. GC separation of TBDMS Derivatized Amino Acids (see Table 9.1) (1 = alanine, 2 = glycine, 3 = valine, 4 = leucine, 5 = isoleucine, 6 = proline, 7 = methionine, 8 = serine, 9 = threonine, 10 = phenylalanine, 11 = aspartic acid, 12 = hydroxyproline, 13 = cycsteine, 14 = glutamic acid, 15 = asparagine, 16 = lysine, 17 = glutamine (peak 1), 18 = arginine, 19 = histidine, 20 = glutamine (peak 2), 21 = tyrosine, 22 = tryptophan, and 23 = cystine.

80 (3 min)–190° at 4°/min. Separates *d*- and *l*-isomers of AAs (see Figure 9.2).

II. Derivatization of Amino Acids and PTH-Amino Acids

A. *t*-Butyldimethylsilyl (TBDMS) Derivative[1]

Reagents: *N*-methyl-*N*-(*tert*-butyldimethylsilyl), trifluoroacetamide, MTBSTFA: Add 0.25 ml of *N,N*-dimethylformamide (DMF) to the dried hydrolyzate. Add 0.25 ml of MTBSTFA reagent and cap tightly. Heat at 60° for 60 min, or overnight at room temperature (longer reaction times prevent mixtures of derivatives). Sample will need to be concentrated prior to injection. For trace analyses, it is important to use the minimum amount of solvent.

B. *N*-PFP Isopropyl Ester Derivative

Evaporate the sample to dryness with clean, dry nitrogen. Add 0.5 ml of 2N HCl in isopropyl alcohol. Heat at 100° for 1 hour. If tryptophan and/or cystine are suspected of being present, add 1 ml of ethyl mercaptan to prevent oxidation. Evaporate the reaction

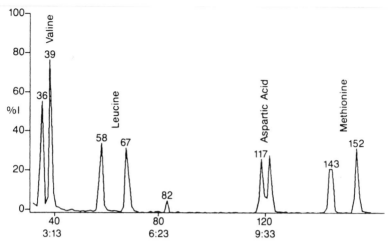

Figure 9.2.   GC Separation of *N*-PFP Isopropylesters of D and L Amino Acids

mixture to dryness with clean, dry nitrogen. Add 0.5 ml ethyl acetate
and 50 μl of pentafluoropropionic anhydride (PFPA). Heat at 100°
for 30 min. Evaporate again with clean, dry nitrogen. Dissolve
residue in the minimum amount of methylene chloride.

$$NH_2CH(R)COOH$$

$$\xrightarrow[\text{2. PFPA}]{\text{1. } (CH_3)_2 \text{ CHOH}} C_2F_5C(O)NHCH(R)C(O)OCH(CH_3)_2$$

$$MW = 262 + R$$

This procedure works well with alanine, valine, threonine, isoleucine, glycine, leucine, proline, serine, aspartic acid, cystine, methionine, phenylalanine, tyrosine, ornithine, and lysine.

III. Mass Spectral Interpretation

A. Mass Spectra of TBDMS Derivatives of Amino Acids
   Typical fragment ions are M − 15 (CH_3), M − 57 (C_4H_9), M − 85
   (C_4H_9 + CO), M − 159 (C(O)−O−TBDMS).

   In general, if two fragment ions are observed that are 28 mass
   units apart, then 57 (C_4H_9) is added to the highest fragment ion to
   deduce the molecular weight of the TBDMS derivative.

   Identification of unknowns, as well as confirming the presence
   of known AAs, is more reliable if accurate mass measurement

Table 9.2.   Characteristic ions of TBDMS derivitized amino acids

Deduced MW	*m/z* Values of other ions	Amino acid
303	218, 246	Glycine
317	158, 232, 260	Alanine
343	184, 258, 286	Proline
345	186, 260, 288	Valine
359	200, 274 > 302	Leucine (elutes first)
	200, 274 < 302	Isoleucine
377	218, 292, 320	Methionine
393	234, 302, 336	Phenylalanine
447	362, 390	Serine
461	303, 376, 404	Threonine
463	378, 406	Cysteine
473	314, 388, 416	Hydroxyproline
474	302, 417	Asparagine
475	316, 390, 418	Aspartic acid
488	300, 329, 431	Lysine (elutes first)
	299, 329, 357, 431	Glutamine (first of two peaks)
489	272, 330, 432	Glutamic acid
497	196, 338, 440, 459	Histidine
499	199, 340, 442	Arginine (first of two peaks)
523	302, 364, 438, 466	Tyrosine
546	244, 302, 489	Tryptophan
696	348, 537, 589, 639	Cystine

data are also available. Identification is easily accomplished if mass spectra of the AAs are added to the computer-assisted library search routine (see Table 9.2). Selected ion monitoring of characteristic ions using previously determined retention time windows is generally used for trace analyses (see Table 9.1 and Figure 9.1). Accurate mass SIM reduces chemical noise at the expense of transmitted ion current.

B. Mass Spectra of Underivatized PTH-Amino Acids
  The phenylthiohydantion (PTH) derivative plus the AA backbone has a mass of 191 Daltons. Thus, subtracting 191 from the molecular ion of the PTH-AA derivative gives the mass of the AA side chain. A characteristic fragment ion is observed at *m/z* 135. Leucine and

isoleucine can be differentiated in the mass spectra by the absence of m/z 205 in the isoleucine spectrum.

C. Mass Spectra of TBDMS Derivatives of PTH-Amino Acids
   Multiple TBDMS derivatives may form depending on the R group of the AA and the reaction time. Some of the PTH-AAs form TBDMS derivatives in less than 1 hour while others require overnight reaction times.

The MW of the TBDMS derivatives of PTH-AAs can be calculated using the following formula: MW $= R + 191 + n(114)$, where n is the number of TBDMS groups. Arginine loses NH_3 during, or prior to, the derivatization reaction. Thus, the characteristic loss of 57 Daltons (C_4H_9) from the TBDMS derivative occurs at M $-$ 74 (57 + 17).

   If the derivatization does not go to completion, it is a good idea to plot the values given in Table 9.3 and the values that are 114 Daltons less to determine the presence of a particular PTH-AA.

D. Mass Spectra of *N*-PFP Isopropyl Esters of D and L Amino Acids
   The molecular ion is usually not observed but can be deduced by adding 87 ($COOC_3H_7$) Daltons to the most abundant, high-mass ion. Masses 69 (CF_3) and 119 (C_2F_5) may also be observed.[2]

E. Sample Mass Spectrum TBDMS Derivatized Amino Acids
   Examining the mass spectrum of glutamic acid-TBDMS shows two high-mass ions that are 28 mass units apart (see Figure 9.3). By adding 57 (C_4H_9) Daltons to the m/z 432 ion, the MW of the

Table 9.3.   Selected ion monitoring of some PTH-AAs as the TBDMS derivatives

Component	Suggested ions to monitor
PTH-alanine-TBDMS	377, 734
PTH-glycine-TBDMS	363, 420
PTH-valine-TBDMS	291, 405, 462
PTH-leucine-TBDMS	419, 433, 476
PTH-isoleucine-TBDMS	419, 447, 476
PTH-proline-TBDMS	232, 346
PTH-methionine-TBDMS	437, 494
PTH-serine-TBDMS	507, 564
PTH-threonine-TBDMS	521, 578
PTH-phenylalanine-TBDMS	453, 510
PTH-aspartic acid-TBDMS	507, 535, 592
PTH-cysteine-TBDMS	—
PTH-glutamic acid-TBDMS	549, 606
PTH-asparagine-TBDMS	534, 591
PTH-lysine-TBDMS	548, 605
PTH-glutamine-TBDMS	416, 446, 473, 662 (MW 719)
PTH-arginine-TBDMS	559, 616
PTH-histidine-TBDMS	557, 614
PTH-tyrosine-TBDMS	563, 640
PTH-tryptophan-TBDMS	606, 663
PTH-cystine-TBDMS	—
PTH-*S*-carboxymethylcysteine-TBDMS	353, 410

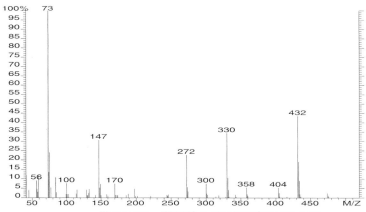

Figure 9.3.   TBDMS Derivatized Glutamic Acid

TBDMS derivative is deduced to be 489. Characteristic fragment ions of the TBDMS derivative include the following:

M − 15 (CH$_3$)	*m/z* 474
M − 57 (C$_4$H$_9$)	*m/z* 432
M − 85 (C$_4$H$_9$ + CO)	*m/z* 404
M − 159 (C(O) − TBDMS)	*m/z* 330

IV. References

1. Kitson, F. G., and Larsen, B. S. In *Mass Spectrometry of Biological Materials.* C. N. McEwen and B. S. Larsen, Eds. New York: Marcel Dekker, 1990.
2. Gelpi, E., Koenig, W. A., Gilbert, J., and Oro, J. Combined GC-MS of Amino Acid Derivatives. *J. Chromat.*, 7, 604, 1969.

Chapter 10

Common Contaminants

I. Contaminants Occasionally Observed after Derivatization with TMS Reagents

m/z = 73, 75, 201, 117 (MW = 216): octanoic acid-TMS
m/z = 73, 75, 313, 328 (MW = 328): palmitic acid-TMS
m/z = 73, 75, 341, 356 (MW = 356): stearic acid-TMS
m/z = 73, 99, 241, 147, 256 (MW = 256): uracil-TMS
m/z = 73, 75, 111, 147, 275 (MW = 290): adipic acid-TMS
m/z = 73, 117, 147 (MW = 234): lactic acid-TMS
m/z = 73, 130, 45, 59 (MW = 247): aminobutyric acid-DiTMS
m/z = 73, 147, 205 (MW = 308): glycerol-TMS
m/z = 73, 147, 233, 245 (MW = 350): malic acid-TMS
m/z = 73, 178, 135, 193, 192 (MW = 193): benzamide-TMS
m/z = 73, 273, 147 (MW = 480): citric acid-TMS
m/z = 73, 332, 147 (MW = 464): ascorbic acid-TMS
m/z = 75, 73, 67, 55 (MW = 352): linoleic acid-TMS
m/z = 75, 73, 131, 45, 146 (MW = 146): propionic acid-TMS
m/z = 75, 73, 145, 45 (MW = 160): n-butyric acid-TMS
m/z = 75, 73, 159 (MW = 174): valeric acid-TMS
m/z = 75, 117, 45, 43, 73 (MW = 132): acetic acid-TMS
m/z = 91, 165, 135 (MW = 180): benzyl alcohol-TMS
m/z = 105, 206, 73, 308 (MW = 323): benzaminoacetic acid-TMS

m/z = 147, 189, 73 (MW = 204): urea-TMS
m/z = 147, 227, 73, 93 (MW = 242): sulfuric acid-TMS
m/z = 151, 166 (MW = 166): phenol-TMS
m/z = 165, 91, 180, 135 (MW = 180): *o*-cresol-TMS
m/z = 165, 180, 91 (MW = 180): *m*-Cresol-TMS
m/z = 174, 59, 75, 147 (MW = 319): aminobutyric acid-TriTMS
m/z = 179, 105, 77, 135 (MW = 194): benzoic acid-TMS
m/z = 187, 75, 73, 69 (MW = 202): octanol-TMS
m/z = 189, 174 (MW = 189): indole-TMS
m/z = 195, 120, 210 (MW = 252): acetylsalicylic acid-TMS
m/z = 203, 188 (MW = 203): skatole-TMS
m/z = 255, 73, 113, 270, 147 (MW = 270): thymine-TMS
m/z = 266, 281, 192 (MW = 281): aminobenzoic acid-TMS
m/z = 267, 73, 193, 223 (MW = 282): hydroxybenzoic acid-TMS
m/z = 285, 117, 132, 145 (MW = 300): tetradecanoic acid-TMS
m/z = 299, 314, 73 (MW = 314): phosphoric acid-TMS

II. Contaminants Occasionally Observed in Underivatized Samples

Ions Observed	Compound
m/z = 84, 133, 42, 162, 161	Nicotine
m/z = 98, 112, 30, 129	BHMT
m/z = 99, 155, 211	Tributyl phosphate
m/z = 122, 105, 77	Benzoic acid
m/z = 149, 167, 279	Dioctylphthalate
m/z = 194, 109, 55, 67, 82	Caffeine
m/z = 205, 220, 57	Di-*tert*-butyl cresol
m/z = 221, 57, 236, 41, 91	Ionol 100
m/z = 225, 93, 66, 65, 39	Tinuvin-P
m/z = 530, 57, 43, 515, 219	Irganox 1076

III. Column Bleed

GC column bleed is a frequently encountered contaminant of mass spectra when high column temperatures are employed. Modern data systems offer the best way to eliminate this type of contamination by subtracting a spectrum showing column bleed from all other spectra in the GC/MS run.

C h a p t e r   1 1

Drugs and
Their Metabolites

Because drugs and metabolites are typically polar and thermally labile molecules, liquid chromatography/mass spectrometry (LC/MS) rather than GC/MS may be a more desirable approach. However, if GC/MS is used, more structural information may be obtained, particularly using accurate mass measurement electron impact ionization (ei) and chemical ionization (ci) combined with derivatization. It is our experience that wide bore (0.53 mm) GC columns can be used in place of narrow bore (0.25 mm) GC columns, resulting in greater sample capacity and less adsorption. The disadvantages are lower GC resolution and the need to use a jet separator. The splitter arrangement shown in Figure 11.1 should be considered.

I. GC Separations*

  A. Underivatized

   1. Basic drug screen
      30 m DB-1 column, 100–290° at 6°/min or 15 m DB-1301 column, 150–250° at 15°/min (injection port at 280°).

   2. Nicotine, pentobarbital, secobarbital, caffeine, oxazepam, and

---

*For metabolite work, first test the separation on the precursor drug or chemical.

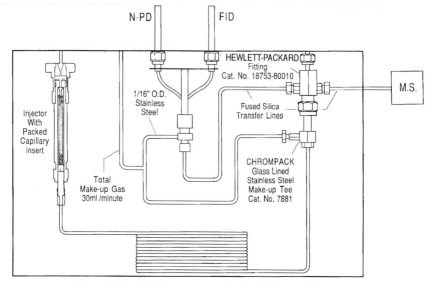

Figure 11.1. Splitter arrangement for wide bore GC columns.

diazepam redissolved in methanol
15 m DB-5 or HP-5 column, 150–300° at 10°/min.

3. Cocaine (MW = 303), codeine (MW = 299) morphine (MW = 285)
25 m DB-1 column, 100–280° at 15°/min.

4. Cocaine metabolites (e.g., ecogonine, benzoylecogonine, etc.)
30 m DB-5 column, 200–280° at 10°/min. (Preparation of methyl ester and TMS derivatives is recommended.)

5. Naloxone (MW = 327) and nalbuphine (MW = 357)
30 m DB-1 column, 150–280° at 15°/min.

6. Amphetamines
30 m DB-1 column at 150°.

7. Fentanyl
30 m DB-1701 column at 270°.

8. Hexamethylphosphoramide and its metabolites
30 m FFAP-DB or DB-WAX column, 60–220° at 10°/min.

9. *o*-Phenylenediamine (OPD) metabolites
   30 m DB-225 column, 75–225° at 10°/min.
   Urine extracts from rats exposed to OPD were examined without
   derivatization. The major metabolites were identified as methyl-
   benzimidazole, methylquinoxaline, and dimethylquinoxaline.

B. TMS Derivatives

1. Naloxone-TMS (MW = 471), nalbuphine-TMS (MW = 573)
   30 m DB-1 column, 60–270° at 10°/min.

2. Methadone metabolites, methadone, cocaine (underivatized),
   morphine, and heroin
   30 m DB-1 column, 100–250° at 10°/min.

3. Daidzein and its metabolites (as TMS derivatives)

Daidzein-TMS	Major Metabolite
$C_{21}H_{26}O_4Si_2$ (MW = 398.1369)	$C_{24}H_{34}O_5Si_3$ (MW = 486.1714)

30 m DB-1 column, 100–280° at 10°/min

C. Acetate Derivatives from Acid Hydrolysis
   Injection port: 280°, 30 m DB-1 column, 100–300° at 10°/min.

D. Methylated Barbituates and Sedatives
   15 or 30 m DB-17 column, 100–250° at 10°/min.

II. Sample Preparation[1–3]

A. Extraction with Solvents
   Drugs and metabolites can be extracted from cultures and urine
   by adding 2 drops of concentrated HCl to 1 ml of urine for a
   pH of 1–2. Extract with three 1-ml volumes of diethyl ether
   (top layer) or methylene chloride (bottom layer). Combine extractions and evaporate with clean, dry nitrogen. Adjust to a pH of
   8–10 by adding 250 μl of 60% KOH to 1 ml of urine. Extract

with three 1-ml volumes of diethyl ether (top layer) or three 1 ml volumes of methylene chloride (bottom layer). Combine extractions and evaporate to dryness with clean, dry nitrogen. (See also reference 2.)

B. Solid-Phase Extraction

1. Prepare the solid-phase extraction (SPE) tube (1 ml LC-18 SPE tube) by conditioning with 1 ml of methanol followed by 1 ml of water.

2. Extract the drugs and metabolites by diluting 1 ml of serum with 1 ml of 0.1M sodium carbonate buffer (pH of 9). Force the mixture dropwise through the SPE tube previously prepared.

   Wash the SPE tube packing with three 200-μl aliquots of water, dry it with nitrogen for 5 min, and elute the drugs with three 100-μl aliquots of 90 parts ethanol and 10 parts diethyl ether. Concentrate the recovered drugs by evaporating some or all of the solvent before analysis by GC/MS. (See Supelco Bulletin 810B.)

3. Clean or change injection port liners frequently because nonvolatile materials in extracts from body fluids can accumulate in the injection port and/or head of the GC column and cause separation problems.

III. Derivatization of Drugs and Metabolites[4]

A. TMS Derivatives of Drugs and Their Metabolites
   Add 100 μl of MSTFA reagent to less than 1 mg of dry extract. Heat at 60° for 15–20 min. If necessary, add 250 μl of acetonitrile or other suitable solvent. For additional structural information, prepare the methoxime-TMS derivative to determine if one or more carbonyl groups are present.

B. MO-TMS Derivative of Drugs
   Add 250 μl of methoxime hydrochloride in pyridine (MOX) reagent to less than 1 mg of dry extract. Let this solution stand at room temperature for 2 hours. Evaporate to dryness with clean, dry nitrogen. Add 250 μl of MSTFA reagent and let stand for 2 hours at room temperature.

C. Acetyl Derivatives of Drugs

Add 60 μl of acetic anhydride and 40 μl of pyridine to less than 1 mg of dry extract. Heat for 1 hour at 60°. Add excess methanol and evaporate to dryness with clean, dry nitrogen. Dissolve the residue in the minimum amount of butyl acetate or ethyl acetate.

D. Methylation of Barbiturates and Sedatives

Heating MethElute (Pierce 49300X) with drug-containing extracts from body fluids gives quantitative methylation of barbiturates, sedatives, and so on. Follow the procedure provided by Pierce Chemical Company using the MethElute reagent.

15 or 30 m DB-17 column, 100–250° at 10°/min.

IV. Mass Spectral Interpretation

A. Metabolites

The mass spectra of metabolites will usually follow similar fragmentation pathways to those prevalent in the mass spectra of the precursor molecule. Thus, a knowledge of the possible biotransformations that can lead to metabolites is important. La Du et al.[2] list oxidation, reduction, and hydrolysis reactions that are common to living organisms. Some of the more common biotransformations are listed in the proceeding text with the exception of conjugated metabolites, which have insufficient volatility to be observed by GC/MS and are not considered. If conjugation is expected, it will be necessary to cleave the conjugate by hydrolysis before GC/MS analysis of the metabolite.[1]

1. Side-chain oxidation and hydroxylation of toluene

| *o*-Cresol | Toluene | Benzyl alcohol |

2. Epoxide formation and hydroxylation of benzene

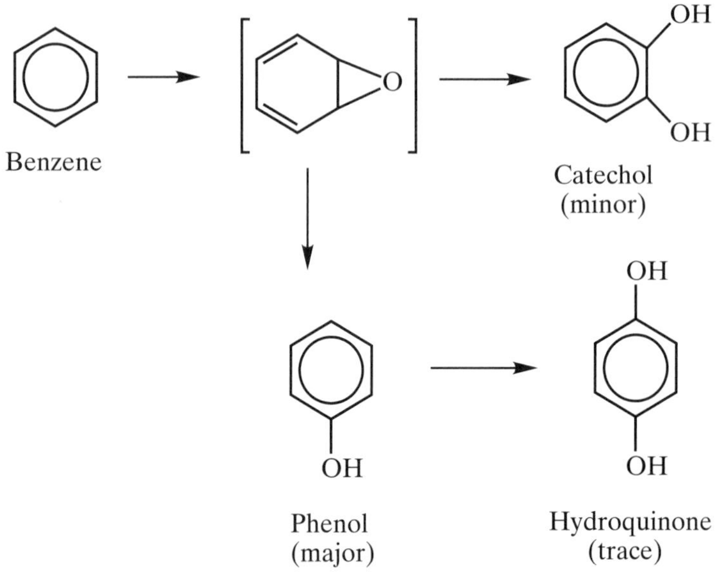

Benzene

Catechol
(minor)

Phenol
(major)

Hydroquinone
(trace)

GC conditions should be used that separate phenol, hydroquinone, resorcinol, and catechol.

3. *o*-Dealkylation of 7-ethoxycoumarin

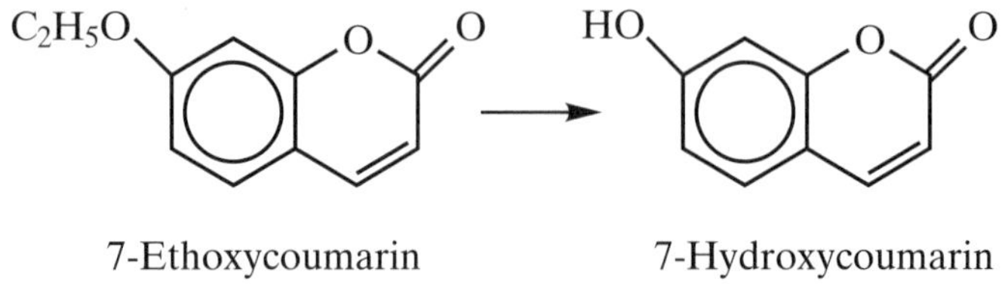

7-Ethoxycoumarin

7-Hydroxycoumarin

4. Hydroxylation and ketone formation of cyclohexane

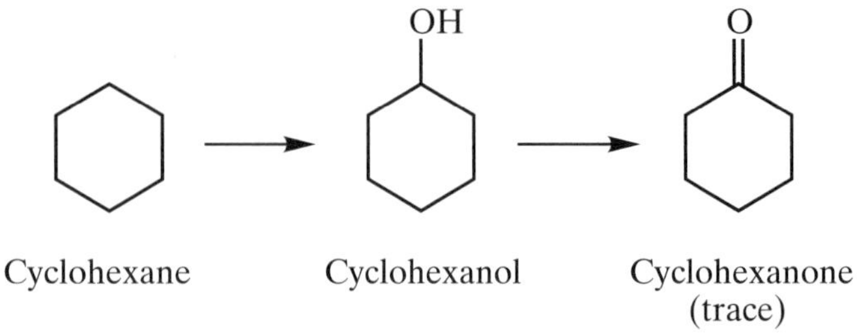

Cyclohexane

Cyclohexanol

Cyclohexanone
(trace)

5. *N*-Dealkylation

$$[(CH_3)_2N]_3P{=}O \rightarrow [(CH_3)_2N]_2P(O)NHCH_3$$
$$\downarrow$$
$$(CH_3)_2N{-}P(O)(NHCH_3)_2$$

By carefully examining the fragmentation pattern of the metabolite and comparison with the mass spectra of the precursor molecule, it is often possible to determine not only the nature of the biotransformation, but also its position in the molecule. In the proceeding example, accurate mass measurement was used to determine that a hydroxyl group had been added to the benzene ring containing the fluorine substituent.

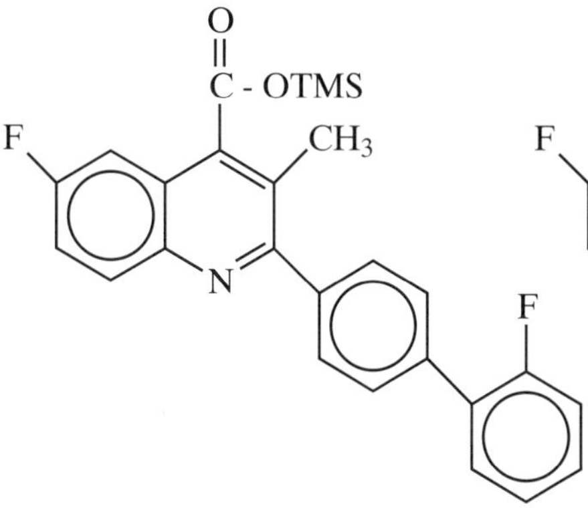

TMS of Precursor
$C_{26}H_{23}O_2F_2N_1Si_1$
(MW = 447.1466)

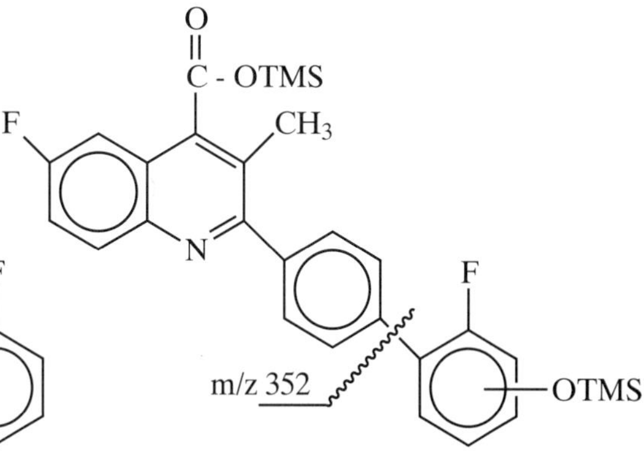

m/z 352

TMS-Major Metabolite
$C_{29}H_{31}O_3N_1F_2Si_2$
(MW = 535.1810)

Molecular ion (MW = 535.1810)	$C_{29}H_{31}O_3NF_2Si_2$
Intense ion (MW = 352.1169)	$C_{20}H_{19}O_2NFSi$
Loss from the molecular ion	$C_9H_{12}OFSi$

The loss of 183.0641 Daltons from the molecular ion showed that the OH group was on the benzene ring containing the fluorine:

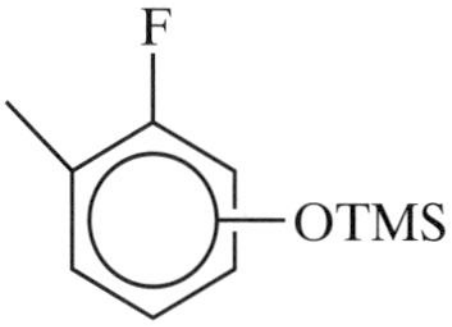

$C_9H_{12}OF_1Si_1$ (MW = 183.0641)

Table 11.1.   Fragmentation and elution order of underivatized drugs*

Drug	Typical fragment ions
Amphetamine (MW = 135)	44, 91, 120
N-methylamphetamine (MW = 149)	58, 91, 134
Nicotine (MW = 162)	84, 133, 162
Ephedrine (MW = 165)	58, 77, 146
Barbital** (MW = 184)	156, 141
Aprobarbital (MW = 210)	162, 124, 195
Tylenol (MW = 151)	109, 151, 80
Phenacetin (MW = 179)	108, 109, 179
Mescaline (MW = 211)	181, 182, 211
Amobarbital** (MW = 226)	156, 141
Pentobarbital** (MW = 226)	156, 141
Meprobamate (MW = 218)	55, 83, 96, 114, 144
Secobarbital (MW = 238)	168, 167, 195
Caffeine (MW = 194)	194, 109, 55
Glutethimide (MW = 217)	189, 117, 132
Hexobarbital (MW = 236)	221, 181, 157, 236
Lidocaine (MW = 234)	86, 58, 72, 234
Phencyclidine (MW = 242)	200, 91, 84
Doxylamine (MW = 270)	58, 71, 183, 182, 200
Theophylline (MW = 180)	180, 95, 68
Phenobarbital (MW = 232)	204, 117, 232
Cyclobarbital (MW = 236)	207, 141
Procaine (MW = 236)	86, 99, 120
Methaqualone (MW = 280)	235, 250, 91
Methadone (MW = 309)	72, 294, 309
Cocaine (MW = 303)	82, 182, 303
Imipramine (MW = 280)	58, 235, 280
Desipramine (MW = 266)	44, 195, 235, 266
Scopolamine (MW = 303)	94, 138, 154, 303
Codeine (MW = 299)	299, 162, 229, 124
Morphine (MW = 285)	285, 162
Chlordiazepoxide (MW = 299)	282, 283, 284
Heroin (MW = 369)	327, 369, 268
Flurazepam (MW = 387)	86, 387, 315
Papaverine (MW = 339)	339, 338, 324
Hydroxyzine (MW = 374)	201, 299, 374
Thioridazine (MW = 370)	98, 70, 370

*GC conditions given in Section I,A,1.
**These drugs can be differentiated by retention time and the m/z 156 and m/z 157 abundance ratios.

B. Drugs

The mass spectra of drugs are as varied as the molecules from which they are formed (see Table 11.1). Two major sources that are available for identifying drugs are computer library search routines and *Mass Spectral and GC Data of Drugs, Poisons and Their Metabolites*.[1]

C. Sample Mass Spectrum

The mass spectrum in Figure 11.2 shows a dominant molecular ion at m/z 151. The odd mass suggests the presence of an odd number of nitrogens. The observed loss of 42 Daltons from the molecular ion suggests an *N*-acetylated compound and the significant ion table (see Part III) suggests that m/z 109 is an aminophenol. By acetylating with acetic anhydride and pyridine, the presence of an OH group will be confirmed. The molecular ion for the acetylated material will now be at m/z 193 with corresponding fragments at m/zs 151 and 109.

Another way to derivatize the OH group is by silylation using the Tri-sil Z reagent that will silylate the hydroxyl group, but not silylate the secondary amino group.

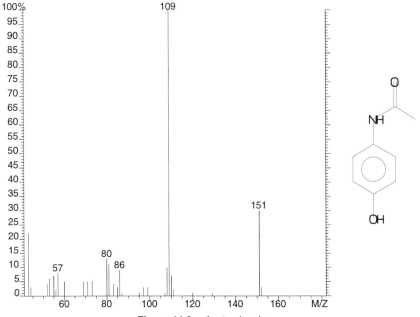

Figure 11.2. Acetaminophen.

References

1. Pfleger, K., Maurer, H., and Weber, A. *Mass Spectral and GC Data of Drugs, Poisons and Their Metabolites.* Weinheim, Germany: VCH Publishers, 1985.
2. LaDu, B. N., Mandel, A.G., and Way, E. L. *Fundamentals of Drug Metabolism and Drug Disposition.* Baltimore: Williams and Wilkins, 1972.
3. Sunshine, I., and Caplis, M., Eds. *CRC Handbook of Mass Spectra of Drugs.* Boca Raton, FL: CRC Press, 1981.
4. Ahuja, S. *J. Pharm. Sci.*, *65*, 163, 1976. (See also *Hewlett Packard Ion Notes*, *6*(2), 1991.)

C h a p t e r   1 2

Esters

I. GC Separation of Esters of Carboxylic Acids

A. Capillary Columns

1. Methyl esters (general)
   30 m FFAP column, 60–200° at 4°/min.

2. C_{14}-C_{24} methyl esters
   30 m DB-23 column, 150–210° at 4°/min.

3. Methyl esters (C_8–C_{20} polyunsaturated)
   30 m DBWAX column, 80–225° at 8°/min or Omegawax 320 column, 100–200° at 10°/min. Run for 50 min.

4. Butyl esters
   30 m DB-1 column, 70–250° at 6°/min.

5. Diesters: dimethyl malonate, dimethyl succinate, dimethyl glutarate, and dimethyl adipate
   30 m CPSIL-88 column, 150–220° at 4°/min or 30 m DB-Wax column, 50–100° at 8°/min, 100–200° at 10°/min.

6. Dimethyl terephthalate (DMT) impurities

Compound	Retention Time (min.)	m/z
Acetone	0.9	58
Benzene	1.3	78
Toluene	2.2	92
Xylene	2.7	106
Methyl benzoate	4.8	150

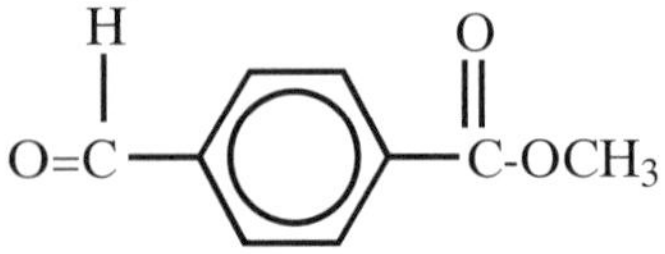

5.9   164

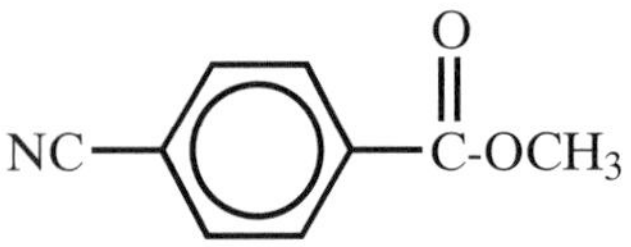

6.0   161

| Methyl toluate | 6.8 | 164 |
| $CH_3C(O)OC_6H_4C(O)OCH_3$ (DMT) | 10.7 | 194 |

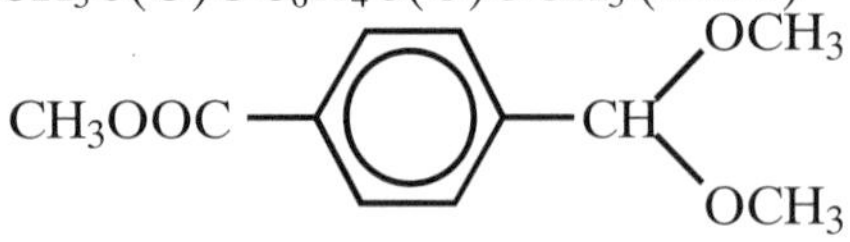

12.3   210

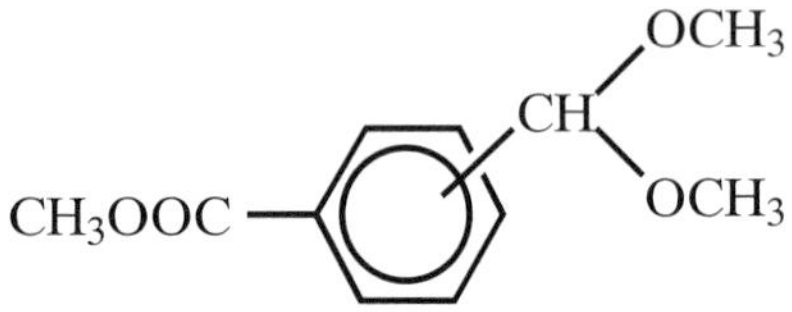

13.6   210

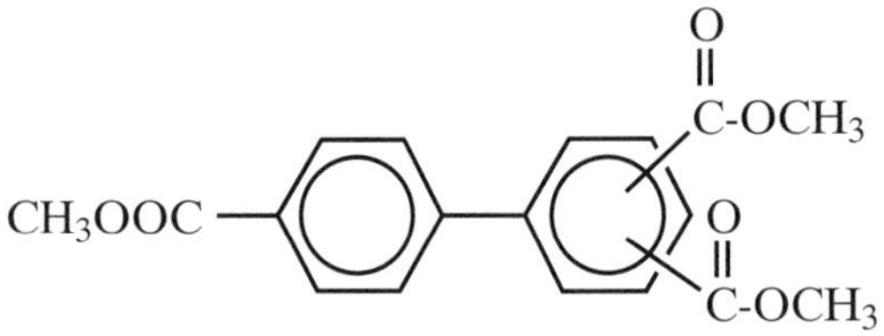

14.5   328

30 m DB-1 column, 100–250° at 10°/min.

7. Glycol methacrylates (and retention times [RTs]): ethylene gly-col dimethacrylate (RT = 13 min), diethyleneglycol dimeth-acrylate (RT = 18 min), triethylene glycol dimethacrylate (RT = 22 min), and tetraethylene glycol dimethacrylate

(RT = 30 min)
30 m DB-17 column, 60–230° at 8°/min.

8. Methyl esters of cyano acids
30 m DB-17 column, 75–275° at 10°/min.

B. Packed Columns

1. Low-boiling methyl esters
2 to 3 m SP-1000 column on Carbopack C column, 75–175° at 8°/min.

II. Mass Spectra of Esters

A. Methyl Esters of Aliphatic Acids

1. General formula: RCO_2Me

2. Molecular ion: If a molecular ion is not observed, it can be deduced by adding 31 or 32 mass units to the highest peak observed (neglecting isotopes).

3. Fragmentation: Cleavage of bonds that are adjacent to the carbonyl group gives rise to R^+, $[RC{\equiv}O]^+$, and $[OCH_3]^+$. An intense peak in the mass spectra of C_6–C_{26} methyl esters results from the McLafferty rearrangement:

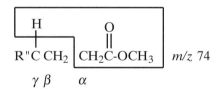

The McLafferty rearrangement results from β-cleavage and the transfer of a γ-hydrogen. Also characteristic of methyl esters is *m/z* 87, which results from γ-cleavage. The combination of *m/z* 74 and 87 suggests a methyl ester. Branching at the α-carbon may give rise to *m/z* 88, 102, 116, and so on, rather than *m/z* 74. These ions are also observed with ethyl and higher esters (see the proceeding text). If even small peaks are observed at *m/z* 31, 45, 59, and so forth, then it is likely that oxygen is present. Accurate mass measurement data would establish the presence of oxygen, especially in the more intense ions. The following list summarizes the mass spectral characteristics for methyl esters:

In summary: R⎨C—⎨OCH₃

with O double-bonded to C above.

- $[M - 31]^+$ (OCH₃)
- $[M - 43]^+$
- $[M - 59]^+$ (C(O)OCH₃)
- R^+ is intense in methyl acetate, methyl propionate, methyl butyrate, and methyl pentanoate.
- m/zs 74 and 87 are characteristic of straight-chain C_4–C_{26} methyl esters.

Unsaturation in an ester is normally apparent from the molecular weight and the more abundant molecular ion.

B. Determination of the Double-Bond Position in C_{10}–C_{24} Monounsaturated Methyl or Ethyl Esters

1. Derivatization: Dissolve less than 1 mg of methyl or ethyl ester in 1 ml of freshly distilled pyrrolidine and 0.1 ml of acetic acid. Heat the mixture in a sealed or capped tube (able to withstand high temperatures) at 100° for 30 min. Cool the reaction mixture to room temperature and add 2 ml of methylene chloride. Wash the methylene chloride extract with dilute HCl, followed by ion exchange water. Dry the methylene chloride extract with anhydrous magnesium sulfate.[1]

2. GC separation conditions for derivatized esters: 30 m DB-1 column at 175°.

3. Mass spectra of pyrrolidine derivatives of methyl or ethyl esters: The molecular ions are intense with characteristic fragment ions at m/zs 70, 98, and 113. The position of the double bond can be determined by locating two peaks that differ by 12 mass units. A relatively intense peak should be observed 26 m/z units higher than the lower mass peak. The double bond lies between the two peaks that are separated by 26 Daltons.
    In Figure 12.1, peaks are observed at m/zs 126, 140, 154, 168, 182, 196, 210, 224, 236, 250, 264, 278, 292, 306, 320, and 335 (MW).

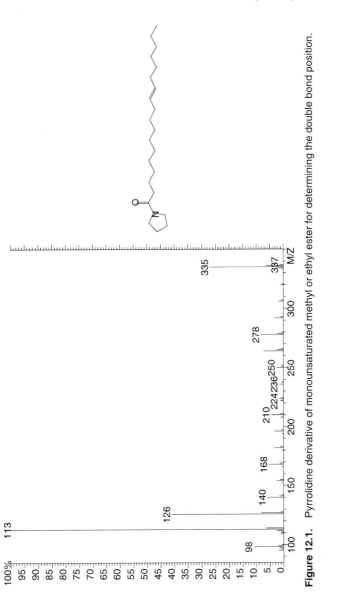

Figure 12.1. Pyrrolidine derivative of monounsaturated methyl or ethyl ester for determining the double bond position.

$$\left[\text{-}N\text{-}\!\!\!\!\left|\overset{\overset{\displaystyle O}{\|}}{\underset{70}{\text{-}C\text{-}}}\right|^{98}\!\!\!(CH_2)_9\left.\right|^{224}\!\!CH\text{=}CH\left.\right|^{250}\!\!C_6H_{13}\left.\right]^{335}$$

This procedure also applies to the case when two double bonds are present with only a CH_2 group separating them. (Not all cases have been examined.)

C. Esters ($R' > C_1$)

The molecular ion can be very small or nonexistent. Esters where R' is greater than methyl form a protonated acid that aids in the interpretation (e.g., m/z 47, formates; m/z 61, acetates; m/z 75, propionates; m/z 89, butyrates; etc.). Interpreting the mass spectra of ethyl esters may be confusing without accurate mass measurement because the loss of C_2H_4 can be confused with the loss of CO from a cyclic ketone.

A summary of the characteristic fragmentation of esters higher than methyl is as follows (see Figure 12.2):

- $M^{+\cdot}$ is small.
- $[M - R']^+$
- $[M - OR']^+$
- $[RCOOH_2]^+$
- $[R - H]^+$ or $[R - 2H]^+$
- McLafferty rearrangement m/z 88 for ethyl esters: If the α-carbon has a methyl substitution, the major rearrangement ion will be at m/z 102.

D. Methyl Esters of Dibasic Acids

Molecular ions are not always observed for methyl esters of dibasic acids; however, methyl esters of all dibasic acids lose 31 and 73 Daltons from their molecular ions. By adding 31 or 73 to the appropriate ions, the molecular weight can be deduced.

E. Ethyl and Higher Esters of Dibasic Acids

1. General formula: $ROOC(CH_2)_nCOOR'$

2. Molecular ion: The molecular ion is generally not observed for diesters larger than diethyl malonate.

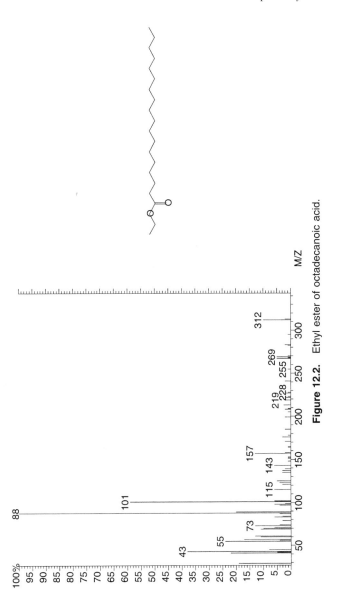

Figure 12.2. Ethyl ester of octadecanoic acid.

3. Fragmentation: Characteristic losses from the molecular ions are as follows:

- $[M - (R' - H)]^+$
- $[M - (OR')]^+$
- $[M - COOR]^+$

F. Aromatic Methyl Esters

1. General formula: $ArC - OCH_3$

2. Molecular ion: Although the molecular ion is always observed, the loss of 31 Daltons (OCH_3) is the most intense ion. Generally, the acid and/or protonated acid is observed. Ortho substituents are distinguished by their large peaks at $[M - 32]^+$, as well as $[M - 31]^+$. A small peak is observed at $[M - 60]^+$.

3. Fragmentation: Characteristic ions are

- $M^{\cdot+}$
- $[M - 31]^+$ and/or $[M - 32]^+$ (ortho effect)
- $[M - COOCH_3]^+$
- $[M - HCOOCH_3]^+$ (May be small unless an ortho group is present; however, the ortho group need not be a methyl group.)

G. Ethyl or Higher Aliphatic Esters of Aromatic Acids

1. General formula: $ArC - OR$

2. Molecular ion: The molecular ion is observed when $R \leq C_4$. When $R \geq C_5$, the molecular ion is very small or nonexistent.

3. Fragmentation: The most characteristic fragment of higher esters is the loss of the alkyl group (less one or two hydrogen atoms) through rearrangement. Characteristic ions are:

- $[M]^+$ (small or nonexistent)
- $[M - (R-H)]^+$ and $[M - (R-2H)]^+$
- $[ArCOO]^{+\cdot}$ or $[ArCOO + 2H]^+$

Generally the acid or protonated acid is observed. The aromatic alcohols can be differentiated by the loss of 18 Daltons from the molecular ion.

H. Methyl Esters of Cinnamic Acid

1. General formula:

$$Ph-CH=CH-\overset{\overset{\displaystyle O}{\|}}{C}-OCH_3$$

2. Molecular ion: The molecular ion is abundant.

3. Fragmentation: The esters of cinnamic acid follow the fragmentation of benzoates except there is a relatively abundant M − 1 peak. Prominent ions are as follows:

- $[M]^{+\cdot}$
- $[M - 1]^+$
- $[M - 31]^+$
- m/z 103
- m/z 77

I. Methyl Esters of Benzene Sulfonic Acids

Example 12.1

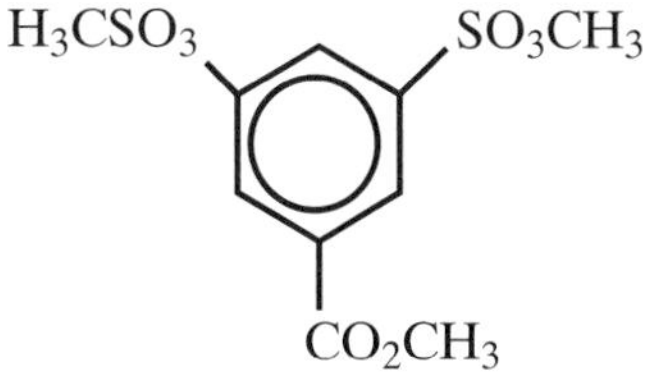

H_3CSO_3 ⟨benzene ring⟩ SO_3CH_3 / CO_2CH_3

Prominent ions are as follows:

- $[M]^{+\cdot}$
- $[M - 31]^+$ (OCH_3)
- $[M - 95]^+$ (SO_3CH_3)
- $[M - 126]^+$ $(SO_3CH_3 + CH_3O)$
- $[M - 154]^+$ $(SO_3CH_3 + CO_2CH_3)$
- $[M - 190]^+$ $(SO_3CH_3 + SO_3CH_3)$

Example 12.2

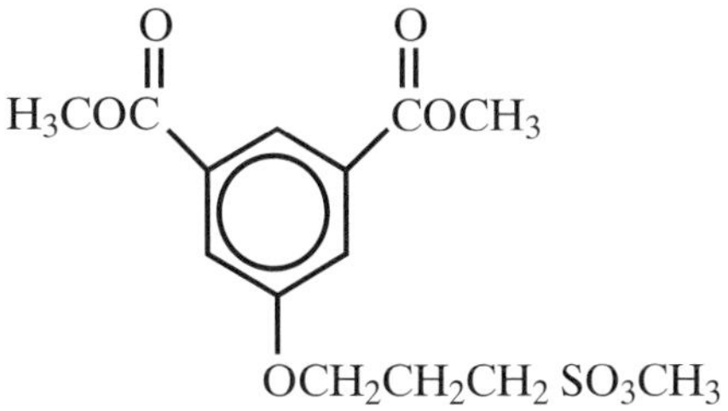

H_3COC ⟨benzene ring⟩ $COCH_3$ / $OCH_2CH_2CH_2\,SO_3CH_3$

Prominent ions are as follows:

- $[M]^{\cdot+}$
- $[M - 31]^+$ (OCH_3)
- $[M - 136]^+$ ($CH_2{=}CHCH_2SO_3CH_3$)
- *m/z* 193 ($M - OCH_2CH_2CH_2SO_3CH_3$)
- *m/z* 137 [(CH_2)$_3SO_3CH_3$]
- *m/z* 95 (SO_3CH_3)

J. Phthalates (see Chapter 28)

K. Esters of Cyano Acids

The molecular ion typically is not observed but can be deduced by adding 31 to the highest-mass peak observed (even in the case of methyl cyanoacetate). An intense ion in the mass spectra of methyl esters is *m/z* 74, formed in a McLafferty rearrangement. Also present is *m/z* 87, but it is of low intensity. Characteristic losses from the molecular ion are 31, 40 (small), 59, and 73 Daltons.

L. Methyl Esters of Aromatic Cyano Acids

The methyl esters of aromatic cyano acids show intense molecular ions, but the intensity decreases as the length of the side chain increases. Losses from the molecular ion are 31 and 59 Daltons.

Reference

1. Anderson, B. A., and Holman, R. T. *Lipids, 9,* 185–190, 1974.

Chapter 13

Ethers

I. GC Separation of Ethers

A. Capillary Columns

1. Epichlorohydrin, propylglycidyl ether, allylglycidyl ether, phenoxy-2-propanone, and phenylglycidyl ether
30 m DB-1 column, 60–175° at 8°/min.

2. Benzoin methyl ether, benzoin propyl ether, benzoin isopropyl ether, and benzoin isobutyl ether
30 m DB-1 column, 150–250° at 10°/min.

3. Ethylene oxide, 2-chloroethanol, and ethylene glycol in DMF
30 m DB-Wax column, 60 (2 min)–180° at 15°/min.

4.

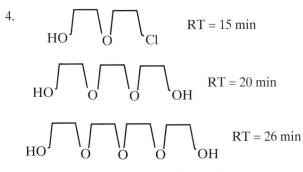

30 m DB-17 column, 75–250° at 6°/min.

117

5. Dowtherm (biphenyl + biphenyl ether) and its impurities (phenol, dibenzofuran, benzophenone, terphenyl, phenoxybiphenyl, xanthone, and diphenoxybenzene)
   30 m DB-225 column, 75–215° at 8°/min. Run for 40 min.

B. Packed Columns

1. Diisopropyl ether and di-*n*-propyl ether
   3 m Tenax-GC column, 140–300° at 15°/min.

2. Ethylene oxide and ethanol
   3 m Porapak QS column, 50–125° at 10°/min.

3. Butyl alcohol, ethylene glycol monoethyl ether, cyclohexanol, benzyl alcohol, diethylene glycol monobutyl ether, and dibutyl phthalate
   3 m packed Tenax-GC column, 100–300° at 10°/min.

4. THF and its impurities (acetone, acrolein, 2,3-dihydrofuran, butyraldehyde, isopropyl alcohol, tetrahydrofuran, 1,3-dioxolane, 2-methyltetrahydrofuran, benzene, and 3-methyltetrahydrofuran)
   3 m 3% SP-1000 column on Carbopack B, 60–200° at 6°/min.

II. Mass Spectra of Ethers

A. Unbranched Aliphatic Ethers

1. General formula: ROR′

2. Molecular ion: The molecular ion intensity decreases with increasing molecular weight, but is still detectable through C_{16} although the ion abundance is low (0.1%).

3. Fragmentation: In general, aliphatic ethers undergo α-cleavage to the oxygen when symmetrical, but undergo β-cleavage when unsymmetrical.

Example 13.1

$$C_3H_7 \overset{\overline{59}}{\underset{}{\Big\{}} CH_2OC_2H_5$$

The β-cleavage is usually in the longer chain.
The presence of peaks at *m/z*s 31, 45, 59, 73, and so on, indicates the presence of oxygen, and these peaks are usually formed by

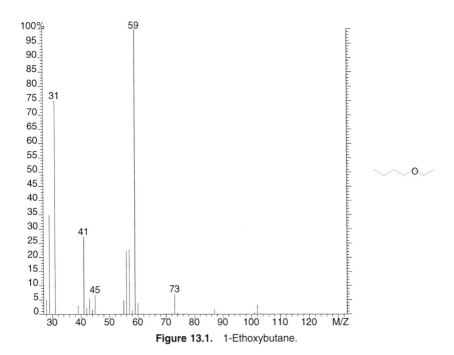

Figure 13.1. 1-Ethoxybutane.

a rearrangement process. The lack of a peak corresponding to the loss of water helps distinguish ethers from alcohols.

B. Cyclic Ethers

1. General formula:

$$H_2C \overline{\hspace{2cm}} (CH_2)_n$$

$$\diagdown O \diagup$$

2. Molecular ion: Molecular ion is generally present and occurs at m/zs 44, 58, 72, 86, 100, 114, and so forth.

3. Fragmentation: Losses from the molecular ions are 1, 29, and 30 Daltons with the loss of 29 (CHO) being characteristic of cyclic ethers. This loss also appears in the mass spectra of unsaturated cyclic ethers, such as furans and benzofurans. The fragmentation of saturated cyclic ethers generally shows a M − 1 ion.

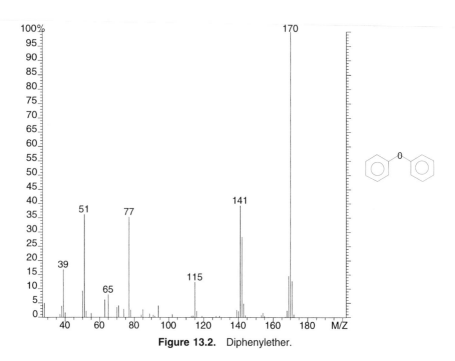

Figure 13.2.   Diphenylether.

C. Diaryl Ethers

1. General formula: ArOAr'

2. Molecular ion: As is expected, the molecular ion peak is intense for aryl ethers.

3. Fragmentation: Losses of 1(H), 28 (CO), and 29 (CHO) Daltons are commonly observed for diaryl ethers. In diphenyl ether, m/z 77 is intense, while in phenyl toluoyl ethers m/z 91 is intense. An intense peak in the mass spectra of alkyl phenyl ethers occurs at m/z 94.

D. Benzoin Alkyl Ethers

1. General formula: PhCH(OR)COPh

2. Molecular ion: The molecular ion intensity is usually small.

3. Fragmentation: Loss of 105 Daltons from the molecular ion is usually observed.

E. Sample Mass Spectra

1. In Figure 13.1, the peaks observed at m/zs 31, 45, 59, and 73 suggest the presence of oxygen in the molecule. By comparing these ions with those listed in Part III, it is evident that alcohols, ethers, and ketones are probable structures. Since no M − 18 peak is observed, the alcohol structure is eliminated. The ion at m/z 59 indicates that the structure contains $C_2H_5OCH_2-$. The difference between the molecular ion observed at m/z 102 and the prominent fragment at m/z 59 is 43 Daltons, which correlates to either CH_3CO or C_3H_7. Hence, the structure for the molecule is $C_2H_5OC_4H_9$.

2. In Figure 13.2, the intensity of the ion at m/z 170 represents a molecular ion of an aromatic compound. The characteristic losses from the molecular ion (M − 1, M − 28, and M − 29) suggest an aromatic aldehyde, phenol, or aryl ether. The molecular formula of $C_{12}H_{10}O$ is suggested by the molecular ion at m/z 170, which can be either a biphenyl ether or a phenylphenol. The simplest test to confirm the structure is to prepare a TMS derivative, even though m/z 77 strongly indicates the diaryl ether.

Chapter 14

Fluorinated Compounds*

I. GC Separations

A. Low-Boiling Fluorinated Compounds, C_1–C_2

Although capillary columns are generally preferred for most applications, packed and porous layer open tubular (plot) GC columns provide the best separation of low-boiling fluorinated compounds.

1. Capillary columns

    a. CF_3Cl, CF_3CF_2Cl, CF_2Cl_2, and CHF_2Cl
    50 m Plot Alumina column (Chrompack cat. no. 4515, 0.32 mm, or cat. no. 7518, 0.53 mm), 60–200° at 5°/min.

    b. SiF_4, CH_3F, CH_3Cl, and C_2H_5F
    30 m GSQ column (J&W Scientific cat. no. 115-3432), 60–150° at 5°/min.

---

*Fluorinated compounds are frequently referred to by code, such as F-115. To translate this code into a molecular formula, add 90 to 115. The first digit of the sum is the number of carbons; the second, the number of hydrogens; the third, the number of fluorines; chlorines complete the valences (i.e., C_2F_5Cl is F-115 = 90 + 115 = 205, or $C_2H_0F_5$). A four-digit number is used for unsaturated molecules. See Tables 14.1–14.3 for the numbering system for chlorofluorocarbons.

Table 14.1.   Numbering system for C_1 and C_2 chlorofluorocarbons*

Industrial number	Formula	Approximate boiling point (°C)	MW
11	$CFCl_3$	24	136
12	CF_2Cl_2	−30	120
13	CF_3Cl	−81	104
14	CF_4	−128	88
20	$CHCl_3$	61	118
21	$CHFCl_2$	9	102
22	CHF_2Cl	−41	86
23	CHF_3	−82	70
30	CH_2Cl_2	40	84
31	CH_2FCl	−9	68
32	CH_2F_2	−52	50
40	CH_3Cl	−24	50
41	CH_3F	−79	34
110	CCl_3CCl_3	185	234
111	$CFCl_2CCl_3$	137	218
112	$CFCl_2CFCl_2$	93	202
113	$CFCl_2CF_2Cl$	48	186
113a	CF_3CCl_3	46	186
114	CF_2ClCF_2Cl	4	170
114a	CF_3CFCl_2	4	170
115	CF_3CF_2Cl	−39	154
116	CF_3CF_3	−78	138
120	$CHCl_2CCl_3$	162	200
121	$CHCl_2CFCl_2$	117	184
121a	$CHClFCCl_3$	117	184
122	$CHCl_2CF_2Cl$	72	168
122a	$CHClFCFCl_2$	73	168
122b	CHF_2CCl_3	73	168
123	CF_3CHCl_2	27	152
123a	$CF_2ClCHClF$	28	152
123b	CHF_2CFCl_2	28	152
124	CF_3CHClF	−12	136
124a	CHF_2CF_2Cl	−10	136
125	CF_3CHF_2	−49	120
130	$CHCl_2CHCl_2$	146	166
130a	CCl_3CH_2Cl	131	166

Table 14.1. (*continued*)

Industrial number	Formula	Approximate boiling point (°C)	MW
131	$CHCl_2CHClF$	103	150
131a	$CH_2ClCFCl_2$	88	150
131b	CH_2FCCl_3	88	150
132	$CHClFCHClF$	59	134
132a	CHF_2CHCl_2	60	134
132b	CF_2ClCH_2Cl	47	134
132c	$CFCl_2CH_2F$	47	134
133	$CHClFCHF_2$	17	118
133a	CF_3CH_2Cl	6	118
133b	CF_2ClCH_2F	12	118
134	CHF_2CHF_2	−20	102
134a	CF_3CH_2F	−27	102
140	$CHCl_2CH_2Cl$	114	132
140a	CH_3CCl_3	74	132
141	$CHClFCH_2Cl$	76	116
141a	$CHCl_2CH_2F$	?	116
141b	$CFCl_2CH_3$	32	116
142	CHF_2CH_2Cl	35	100
142a	$CHClFCH_2F$	?	100
143a	CF_3CH_3	−48	84
150	CH_2ClCH_2Cl	84	98
150a	CH_3CHCl_2	57	98
151	CH_2FCH_2Cl	53	82
151a	CH_3CHClF	16	82
152	CH_2FCH_2F	25	66
160	CH_3CH_2Cl	13	64
161	CH_3CH_2F	−37	48

*The formula can be derived from the number by adding 90 to the industrial number. Reading the ensuing digits from right to left gives the number of fluorines, hydrogens, carbons, and double bonds. The remaining valance positions are reserved for chlorines.

c. CF_3CHClF impurities
105 m RTX-1 capillary column at room temperature or 10 ft SP-1000 column on Carbopack B (Supelco cat. no. 1-1815 M, for the packing.)

Table 14.2. Numbering system for unsaturated C_2 chlorofluorocarbons

Industrial numbering	Formula	Approximate boiling point (°C)	MW
1110	$CCl_2{=}CCl_2$	121	164
1111	$CFCl{=}CCl_2$	71	148
1112 (trans)	$CFCl{=}CFCl$	22	132
1112 (cis)	$CFCl{=}CFCl$	21	132
1112a	$CF_2{=}CCl_2$	19	132
1113	$CClF{=}CF_2$	−28	116
1114	$CF_2{=}CF_2$	−76	100
1120	$CHCl{=}CCl_2$	88	130
1121	$CHCl{=}CFCl$	35	114
1121a	$CHF{=}CCl_2$	37	114
1122	$CF_2{=}CHCl$	−18	98
1123	$CF_2{=}CHF$	−56	82
1130 (cis)	$CHCl{=}CHCl$	60	96
1130 (trans)	$CHCl{=}CHCl$	48	96
1130a	$CH_2{=}CCl_2$	37	96
1131	$CHF{-}CHCl$	11	80
1131 (cis)	$CHF{=}CHCl$	16	80
1131 (trans)	$CHF{=}CHCl$	−4	80
1131a	$CH_2{=}CClF$	−25	80
1132	$CHF{=}CHF$	−28	64
1132a	$CH_2{=}CF_2$	−82	64
1140	$CH_2{=}CHCl$	−14	62
1141	$CH_2{=}CHF$	−72	46
1150	$CH_2{=}CH_2$	−104	28

d. O_2, N_2, CO, CF_4 (freon 14 or F-14), and CF_3CF_3 (freon 116 or F-116)
30 m GS Molecular Sieve column at room temperature (J & W Scientific cat. no. 93802).

e. O_2, N_2, CO, F-14, and F-116
4 m Molecular Sieve 13x column at room temperature.
30 m GS-Alumina column (J & W Scientific cat. no. 115-3532), 60–200° at 5°/min.

f. CHF_3, CH_2F_2, CF_3CH_3, CHF_2Cl, CF_2Cl_2, CH_2FCl, CHF_2CH_2Cl, and $CHFCl_2$
25 m Poraplot Q column, 60–150° at 5°/min.

Table 14.3. Numbering system for cyclic chlorofluorocarbons

Industrial number	Formula	Approximate boiling point (°C)	MW
C-216	F_2C——CF_2 \ / CF_2	-32	150
C-314	F_2C—CCl_2 | | F_2C—CCl_2	132	264
C-317	F_2C—CCl_2 | | F_2C—CCl_2	60	216
C-318	F_2C—CF_2 | | F_2C—CF_2	-6	200

2. Packed columns

   a. CHF_2Cl, CF_2Cl_2, $CHFCl_2$, and CF_2ClCF_2Cl
     3 m 5% Krytox column on 60-80 mesh Carbopack B (Supelco cat. no. 1-2425), 35 (5 min)–160° at 5°/min.
     3 m 5% Fluorocol column on 60–80 mesh Carbopack B (Supelco cat. no. 1-2425M), 70 (8 min)–180° at 8°/min.

B. Medium-Boiling Fluorinated Compounds (Mostly Liquids)

  1. Capillary columns

   a. Fluoroketones and fluoroethers
     30 m DB-210 column, 50–100° at 5°/min (J & W Scientific cat. no. 122-0233).

   b. C_2F_5Cl, $CFCl_3$, CF_2Cl_2, and $C_2F_3Cl_3$
     30 m DB-624 column, 40 (10 min)–140° at 5°/min (J & W Scientific cat. no. 125-1334).

   c. $CF_2{=}CCl_2$, $CCl_2{=}CCl_2$, CCl_3CFCl_2, $CHCl_2CF_2Cl$, CF_3CHCl_2, F-113, and F-112
     30 m DB-Wax column, 50–200° at 5°/min (J & W Scientific cat. no. 125-7032).

d. F-11 and F-113
30 m DB-Wax column at 50°.

2. Packed columns

a. Freon 214 (MW = 252):
$CF_2ClCCl_2CF_2Cl$, $CF_3CCl_2CFCl_2$, $CCl_3CFClCFCl_2$,
$CF_3CClFCCl_3$, $CFCl_2CF_2CFCl_2$, and $CF_2ClCF_2CCl_3$

Freon 215 (MW = 236):
$CF_3CCl_2CF_2Cl$, $CF_2ClCFClCF_2Cl$, $CF_3CFClCFCl_2$,
$CF_2ClCF_2CFCl_2$, and $CF_3CF_2CCl_3$

3 m 1% SP-1000 column on Carbopack B, 40 (5 min) 100° at
95°/min–200° at 10°/min.

b. CH_3Cl, F-22, CH_3Br, F-12, F-21, CH_3I, CH_2Cl_2, F-114, F-11,
$CHCl_3$, CH_2ClCH_2Cl, and CH_3CCl_3, and F-124 impurities
3 m 1% SP-1000 column on Carbopack B, 40–175° at 60°/min.

C. Higher-Boiling Fluorinated Compounds

1. $C_6F_{13}CH_2CH_2OH$, $C_6F_{13}CH_2CH_2I$, $C_8F_{17}CH_2CH_2OH$,
$C_4F_9CH_2CH_2OC(O)C(CH_3) = CH_2$, and
$C_6F_{13}CH_2CH_2OC(O)C(CH_3) = CH_2$
30 m DB-1 column, 60–175° at 4°/min (J & W Scientific cat. no.
125-1032).

2. C_2–C_{12} perfluoroalkyl iodides
30 m DB-Wax column, 75–180° at 8°/min.

II. Mass Spectra of Fluorinated Compounds

A. Fluorinated and Partially Fluorinated Aliphatic Compounds

The molecular ion is usually not observed in the mass spectra of
aliphatic fluorinated compounds (>ethane). Common losses are F,
HF, or CF_3. Frequently observed ions lie at m/zs 31 (CF), 50 (CF_2),
and 69 (CF_3). If mass 69 is intense, a CF_3 group is present. The
presence of a small m/z 51 peak indicates the presence of carbon,
hydrogen, and fluorine. If a m/z 51 peak is intense, then CHF_2 is
present. The absence of m/z 51 and/or m/z 47 (without chlorine)
suggests a perfluorinated compound.

B. Chlorofluorocarbons

The molecular ion is usually not observed, but may be deduced by
adding 35 Daltons to the highest-mass ion observed. In the mass

spectra of chlorofluorocarbons, the evidence of chlorine (or bro-mine) is obvious from the isotope ratios observed in the fragment ions (see Chapter 17). A small rearrangement ion at mass 85 con-taining chlorine indicates that carbon, fluorine, and chlorine are present, even though CF_2Cl does not exist in the compound. If a m/z 85 peak is intense and the isotope indicates the presence of chlorine, then CF_2Cl is present in the structure. Look up the abun-dant ions in the "structurally significant" ion tables of Part III to aid in structural assignments. Also, refer to Table 14.4 for the abundant ions in listed order of decreasing intensities, as many of these spectra may not be in your library search program.

C. Sample Mass Spectrum of Chlorofluorocarbons

The highest-mass peak observed in Figure 14.1 lies at m/z 201. Based on the isotope pattern, it contains two chlorine atoms. The molecular ion is not observed but can be deduced by adding 35 Daltons to the most abundant isotopic mass (m/z 201). The molecu-lar weight is 236, and the molecule contains three chlorine atoms. The abundance of m/z 101 suggests that $CFCl_2$ is present in the molecule. Notice that the m/z 69 peak is more intense than the m/z 85 peak, indicating that CF_3 is also present in the molecule. Subtracting 69 (CF_3) and 101 ($CFCl_2$) from the deduced molecular weight of 236 shows that the remainder is 66 Daltons. The significant ion table for m/z 66 suggests CFCl. The deduced structure is $CF_3CFClCFCl_2$.

D. Aromatic Fluorine Compounds

Intense molecular ions are observed in the mass spectra of fluori-nated benzene, ethyl benzene, toluene, and xylene. Most fluorinated aromatics lose 19 Daltons from the molecular ion, and some lose 50 Daltons (e.g., CF_2). The chlorofluoroaromatics can easily be identified by examining the isotope ratios in the vicinity of the molecular ion.

E. Perfluorinated Olefins

The molecular ions of perfluorinated olefins are usually observed. The m/z 31 ion is frequently more abundant in fluorinated olefins than in fluorinated saturated compounds.

F. Perfluorinated Acids

Molecular ions are usually not observed with perfluorinated acids, but may be deduced by adding 17(OH) or 45 (CO_2H) Daltons to

Table 14.4.   Fluorinated compounds listed by most abundant ion

Base peak	Four next most abundant peaks				Compound	Highest m/z peak >1%*
15	34	33	14	31	CH$_3$F	34
15	69	47	112	31	CH$_3$OCF=CF$_2$	112
18	45	33	31	61	CH$_2$FC(O)OH	78
27	29	28	64	26	C$_2$H$_5$Cl	64
29	27	51	79	77	CH$_3$CH$_2$CF$_3$	79
29	51	79	31	50	CF$_2$ClCHO	114
29	69	31	100	150	CF$_3$CF$_2$CF$_2$CHO	150
30	69	50	31	64	CF$_3$NO	99
31	29	33	61	69	CF$_3$CH$_2$OH	83
31	29	50	69	51	CF$_3$CF$_2$CH$_2$OH	100
31	51	29	69	49	HOCH$_2$(CF$_2$CF$_2$)$_2$H	183
31	81	100	50	69	CF$_2$=CF$_2$	100
31	82	51	29	113	CHF$_2$CF$_2$CH$_2$OH	113
31	95	69	65	29	C$_6$F$_{13}$CH$_2$CH$_2$OH	364
31	107	53.5	88	57	CF$_2$=CFCN	107
33	51	31	32	52	CH$_2$F$_2$	52
33	69	83	51	31	CF$_3$CH$_2$F	102
33	244	163	64	83	(C$_2$H$_2$F$_3$)$_3$PO$_4$	321
39	57	108	31	38	CF$_2$=CFCH=CH$_2$	108
41	39	95	64	28	CH$_3$C(CF$_3$)=CH$_2$	95
43	69	15	—	—	(CF$_3$)$_2$CHOC(O)CH$_3$	210
44	31	25	43	13	CF≡CH	44
45	18	51	28	44	CF$_3$C(O)OH	97
45	64	44	31	33	CH$_2$=CF$_2$	64
45	69	119	100	169	CF$_3$CF$_2$CF$_2$C(O)OH	169
45	80	82	44	26	CH$_2$=CClF	80
46	45	27	44	26	CH$_2$=CHF	46
47	27	45	26	67	CH$_3$CHClF	82
47	33	27	48	46	CH$_3$CH$_2$F	48
47	46	61	27	41	CH$_3$CHFCH$_3$	62
47	66	28	12	31	COF$_2$	66
47	66	33	—	—	N$_2$F$_2$	66
49	48	11	68	19	BF$_3$	68
49	84	86	51	47	CH$_2$Cl$_2$	84
50	52	15	49	47	CH$_3$Cl	50
51	33	31	52	32	CH$_2$F$_2$	52
51	64	194	196	143	CHF$_2$CF$_2$CH$_2$Br	194
51	65	47	45	27	CH$_3$CHF$_2$	66
51	67	31	50	69	CHF$_2$Cl	86
51	69	101	132	151	H(CF$_2$)$_6$H	233
51	69	101	132	151	H(CF$_2$)$_7$H	283
51	69	113	101	132	CHF$_2$CF$_2$CF$_2$CHF$_2$	151
51	69	201	127	100	CHF$_2$(CF$_2$)$_3$I	328
51	83	33	31	101	CHF$_2$CHF$_2$	102
51	83	33	64	101	CHF$_2$CH$_2$CH$_2$CH$_2$F	113
51	85	69	87	101	CF$_2$ClCF$_2$CF$_2$CHF$_2$	201
51	85	69	87	101	CHF$_2$(CF$_2$)$_3$CF$_2$Cl	251
51	85	69	151	87	CHF$_2$CF$_2$CF$_2$Cl	152
51	99	49	101	64	CHF$_2$CF$_2$CH$_2$Cl	151
51	101	69	31	50	CF$_3$CHF$_2$	119
51	101	85	67	31	CF$_2$ClCHF$_2$	135

Table 14.4. *(continued)*

Base peak	Four next most abundant peaks				Compound	Highest m/z peak >1%*
51	130	132	31	111	CHF$_2$Br	130
52	71	33	19	14	NF$_3$	71
59	60	39	27	57	CH$_2$=CHCH$_2$F	60
59	60	39	33	57	CH$_3$CF=CH$_2$	60
61	96	98	26	63	CHCl=CHCl (trans)	96
61	96	98	63	26	CHCl=CHCl (cis)	96
61	96	98	63	26	CH$_2$=CCl$_2$	96
62	27	49	64	26	CH$_2$ClCH$_2$Cl	98
62	27	64	26	25	CH$_2$=CHCl	62
63	27	65	26	83	CH$_3$CHCl$_2$	98
63	82	51	31	32	CHF=CF$_2$	82
63	82	60	31	64	$\begin{array}{c} F_2C-S \\ \mid\quad\mid \\ S-CF_2 \end{array}$	164
63	82	132	31	50	$\begin{array}{c} F_2C-CF_2 \\ \diagdown\;/ \\ S \end{array}$	132
63	132	82	69	31	CF$_3$C(S)F	132
64	28	29	27	66	CH$_3$CH$_2$Cl	64
64	45	31	33	44	CH$_2$=CF$_2$	64
64	45	44	31	33	CHF=CHF	64
65	45	85	31	64	CH$_3$CF$_2$Cl	85
65	64	45	61	33	CH$_3$CF$_2$CH$_3$	65
67	41	54	82	56	C$_6$H$_{11}$F (Fluorocyclohexane)	102
67	69	31	111	32	CHFClBr	146
67	69	47	35	48	CHFCl$_2$	102
67	69	101	51	117	CF$_3$CHClF	136
67	83	33	51	118	CHClFCHF$_2$	118
67	86	48	69	32	SOF$_2$	86
67	99	83	69	79	CHClFCHClF	134
67	117	85	69	119	CHClFCF$_2$Cl	152
68	33	70	49	46	CH$_2$ClF	68
69	51	31	50	—	CHF$_3$	70
69	31	81	100	47	$\begin{array}{c} F_3CFC\text{ - }CF_2 \\ \diagdown\;/ \\ O \end{array}$	147
69	31	100	50	131	$\begin{array}{c} F_2C-CF_2 \\ \diagdown\;/ \\ CF_2 \end{array}$	150
69	31	119	100	50	CF$_3$CF$_2$CF$_3$	169
69	47	50	31	28	CF$_3$C(O)F	116
69	47	66	28	31	CF$_3$OCF$_3$	69
69	47	112	31	—	CH$_3$OCF=CF$_2$	112
69	50	25	31	34.5	CF$_4$	69
69	50	31	152	44	CF$_3$NFCF$_3$	152
69	51	31	151	100	CF$_3$CF$_2$CHF$_2$	151
69	51	65	77	574	CF$_3$(CF$_2$)$_7$CH$_2$CH$_2$I	574
69	51	82	151	31	CF$_3$CHFCF$_3$	151
69	58	31	108	28	CF$_3$CHFCN	108

(continued)

Table 14.4. (*continued*)

Base peak	Four next most abundant peaks				Compound	Highest m/z peak >1%*
69	64	114	45	95	$CF_3CF=CH_2$	114
69	64	133	31	45	$CF_3CH_2CF_3$	133
69	65	45	33	31	CF_3CH_3	84
69	76	50	31	38	CF_3CN	95
69	70	139	89	51	CF_3SF_3	139
69	85	47	201	119	$CF_3CF_2CCl_3$	201
69	85	147	87	97	$CF_3C(O)CF_2Cl$	147
69	85	50	87	35	CF_3Cl	85
69	85	97	31	50	$CF_3CF_2C(O)CF_2Cl$	197
69	85	116	135	31	$CF_3CFClCF_3$	185
69	89	127	70	51	CF_3SF_5	127
69	95	96	45	46	$CHF_2CF=CH_2$	96
69	97	50	31	147	$CF_3C(O)CF_3$	166
69	99	83	51	33	$CF_3OCH_2CF_2CHF_2$	181
69	101	169	51	150	$C_9H_2F_{18}$ (Dihydro HFP trimer)	414
69	102	63	82	32	CF_3SH	102
69	113	132	31	82	$CF_3CHFCHF_2$	133
69	113	132	82	31	$CHF_2CF_2CHF_2$	133
69	113	201	132	82	$CF_3CHFCF_2CF_3$	201
69	114	264	119	145	$CF_2=N(CF_2)_3CF_3$	264
69	116	147	131	—	$CF_3CF=CClF$	166
69	119	31	47	78	$CF_3CF_2C(O)F$	119
69	119	31	50	19	CF_3CF_3	119
69	119	31	131	100	$CF_3CF_2CF_2CF_3$	219
69	119	131	31	181	$(CF_3)_2CFCF_2CF_3$	269
69	119	169	31	47	$C_6F_{12}O_2$ (HFPO dimer)	285
69	119	169	31	100	$CF_3(CF_2)_3CF_3$	269
69	119	169	97	147	$CF_3CF_2CF_2OCF-C(O)F$	285
69	119	169	131	100	$CF_3(CF_2)_4CF_3$	319
69	119	169	131	219	$CF_3(CF_2)_5CF_3$	369
69	119	293	243	343	C_9F_{18} (HFP Trimer B)	431
69	119	293	343	243	C_9F_{18} (HFP Trimer A)	381
69	127	296	31	100	CF_3CFICF_3	296
69	129	131	148	150	CF_3Br	148
69	129	131	229	231	$(CF_3)_2CFBr$	248
69	131	31	100	31	$CF_3CF=CF_2$	150
69	131	119	169	50	$(CF_3)_2CF(CF_2)_2CF_3$	319
69	131	181	100	93	Perfluorinated methylcyclohexane	281
69	131	181	100	293	$C_{10}F_{17}$-CF_3	493
69	134	46	153	65	$CF_3N=SF_2$	153
69	141	47	15	—	$(CF_3)_2C\overset{O-CH_2}{\underset{O-CH_2}{\diagup\diagdown}}$	180
69	145	76	246	32	$CF_3SC(S)SCF_3$	246
69	163	—	—	—	$CF_3CCl=CCl_2$	198
69	166	147	31	131	$CF_3CCl=CF_2$	166
69	169	100	119	50	$CF_3CF_2CF_3$	169
69	169	119	97	100	$C_9F_{18}O_3$ (HFPO trimer)	351
69	169	127	296	177	$CF_3CF_2CF_2I$	296
69	170	63	82	31	CF_3SCF_3	170

Table 14.4. (*continued*)

Base peak	Four next most abundant peaks				Compound	Highest *m/z* peak >1%*
69	181	93	200	31	$(CF_3)_2C=CF_2$	200
69	181	131	100	31	Perfluorinated dimethylcyclohexane	381
69	181	131	281	93	$(CF_3)_2CFCF_2CF=CF_2$	300
69	181	281	231	93	C_6F_{12} (HFP dimer B)	300
69	197	169	31	100	$(CF_3)_2CFC(O)CF(CF_3)_2$	197
69	202	64	133	114	CF_3SSCF_3	202
69	219	127	177	31	$F(CF_2)_4I$	327
72	46	51	39	27	$CH_2=CFCH=CH_2$	72
77	51	59	104	39	$CF_3(CH_2CHF)_2CF=CH_2$	206
77	78	51	59	65	$CF_3(CHFCH_2)_2CF=CH_2$	142
79	59	29	47	69	$F(CF_2)_8CH_2CH_3$	448
80	45	82	44	26	$CHF=CHCl$	80
81	15	51	101	63	$CHF_2CF_2OCH_3$	132
81	67	45	61	83	$CHFClCH_2Cl$	116
81	83	61	45	101	$CFCl_2CH_3$	101
82	63	32	31	50	CF_2S	82
82	132	84	134	31	$CFCl=CFCl$	132
83	51	69	33	145	$H(CF_2CF_2)_2CH_2F$	163
83	85	47	87	48	$CHCl_3$	118
83	85	69	67	31	CF_3CHCl_2	152
83	85	67	115	87	$CHClFCHCl_2$	150
83	85	69	130	199	$CHCl_2CCl_2CF_3$	199
83	85	87	95	99	$CHCl_2CHCl_2$	166
83	85	133	87	135	$CHCl_2CHF_2$	134
83	102	67	32	44	SO_2F_2	102
83	118	67	120	64	SO_2FCl	118
83	162	164	33	64	CF_3CH_2Br	162
85	69	87	185	31	$CF_3CClFCF_2Cl$	185
85	69	119	87	31	CF_3CF_2Cl	119
85	69	147	119	31	$CF_2ClCF_2CF_2CF_2Cl$	235
85	69	169	87	31	$CF_3CF_2CF_2Cl$	185
85	86	28	33	87	SiF_4	104
85	87	50	31	35	CF_2Cl_2	101
85	87	129	131	31	CF_2ClBr	145
85	87	163	50	31	$CF_2ClC(O)CF_2Cl$	198
85	117	119	47	31	$CF_2ClCF_2CCl_3$	217
85	120	122	101	69	$POClF_2$	120
85	135	87	31	137	CF_2ClCF_2Cl	151
89	70	51	35	32	SF_4	108
93	31	162	74	112	$CF_2=CF-CF=CF_2$	162
93	143	162	69	31	$CF_3C\equiv CCF_3$	162
93	143	270	74	31	$\begin{matrix} F_2C-CF_2 \\ \vert \quad \vert \\ FC=C-I \end{matrix}$	270
93	162	31	112	143	Perfluorocyclobutene	162
93	162	143	31	193	Perfluorocyclopentene	212
94	75	31	69	56	$CF_3\equiv CH$	94
95	27	96	77	51	$CF_3CH=CH_2$	96
95	59	29	97	27	$CH_3CH_2CFCl_2$	130
95	61	69	130	31	$CF_3CCl=CH_2$	130
95	69	—	—	—	$CF_3CH=CHCl$	130

(*continued*)

Table 14.4. (*continued*)

Base peak	Four next most abundant peaks				Compound	Highest m/z peak >1%*
95	77	—	—	—	$CF_3CH=CH_2$	96
95	130	132	97	60	$CHCl=CCl_2$	130
96	70	50	75	95	Fluorobenzene	96
97	83	99	85	61	$CHCl_2CH_2Cl$	132
97	99	61	117	119	CH_3CCl_3	132
98	100	48	63	31	$CHCl=CF_2$	98
99	51	69	79	129	Hexfluoroisopropyl alcohol	149
99	101	49	85	79	CF_2ClCH_2Cl	134
99	129	178	127	101	$CHF_2CHClBr$	178
100	29	31	64	33	$\begin{array}{c} F_2C\!-\!\!-\!\!CF_2 \\ \mid \quad\quad \mid \\ O\!-\!\!-\!\!CH_2 \end{array}$	111
100	51	49	45	64	CHF_2CH_2Cl	100
100	63	82	113	264	$\begin{array}{c} \diagup CF_2CF_2 \diagdown \\ S \quad\quad\quad\quad S \\ \diagdown CF_2CF_2 \diagup \end{array}$	264
100	63	232	113	150	$\begin{array}{c} F_2C\!-\!\!-\!\!CF_2 \diagdown \\ \mid \quad\quad\quad \mid \quad\; S \\ F_2C\!-\!\!-\!\!CF_2 \diagup \end{array}$	232
100	69	231	181	131	Perfluorodimethylcyclobutane	281
100	131	31	69	50	Perfluorocyclobutane	131
100	131	181	69	31	Perfluoromethylcyclobutane	231
101	51	31	111	113	CHF_2CF_2Br	180
101	83	103	85	149	$CFCl_2CHCl_2$	184
101	85	103	31	87	$CF_2ClCFClCF_2CFCl_2$	267
101	103	66	31	167	$CFCl_2C(O)CFCl_2$	195
101	103	66	35	31	$CFCl_3$	117
101	103	167	169	31	$CFCl_2CFCl_2$	167
101	133	103	67	135	$CFCl_2CHClF$	168
101	228	51	82	127	CHF_2CF_2I	228
104	85	69	50	31	POF_3	104
105	86	67	32	107	SOF_4	105
105	376	77	182	165	$\begin{array}{c} O \\ \parallel \\ \quad\quad C \\ \phi\, C \diagdown \;\diagup O \\ \diagdown \quad\quad \diagup \\ O-C(CF_3)_2 \end{array}$	376
109	110	83	57	63	*o*-Fluorotoluene	110
109	244	175	194	31	1,2-Dichlorohexafluorocyclopentene-1	244
111	84	83	57	28	Fluoroaniline	111
111	85	—	—	—	$CHF_2CH=CCl_2$	146
112	64	63	92	83	Fluorophenol	112
112	64	140	125	92	*o*-Fluorophenetole	140
112	140	84	83	29	*p*-Fluorophenetole	140
113	69	31	132	82	$CF_3CH=CF_2$	132
113	69	63	182	31	$CF_3C(S)CF_3$	182
113	69	163	31	182	Fluorobutene	182
114	69	264	119	145	$CF_2=N(CF_2)_3CF_3$	264
114	116	79	44	81	$CHCl=CFCl$	114

Table 14.4. (*continued*)

Base peak	Four next most abundant peaks				Compound	Highest m/z peak >1%*
115	117	101	103	79	CFCl$_2$CH$_2$Cl	115
116	31	66	85	118	CF$_2$=CClF	116
116	31	118	132	93	1,3-Dichlorohexafluorocyclobutane	197
117	119	121	82	47	CCl$_4$	117
117	119	167	165	83	CHCl$_2$CCl$_3$	200
117	119	167	169	47	CF$_2$ClCCl$_3$	167
117	119	201	203	199	CCl$_3$CCl$_3$	199
117	198	196	119	129	CF$_3$CHClBr	196
117	248	246	167	79	Bromopentafluorobenzene	246
118	83	33	49	120	CF$_2$ClCH$_2$F	118
118	83	33	120	49	CF$_3$CH$_2$Cl	118
119	69	129	131	31	CF$_3$CF$_2$Br	198
119	69	30	264	183	C$_2$F$_5$SSC$_2$F$_5$	302
119	246	127	69	177	CF$_3$CF$_2$I	246
126	75	57	76	31	CF$_2$=CHCH=CF$_2$	126
126	83	111	57	95	Fluoroanisole	126
127	89	108	54	129	SF$_6$	127
129	69	164	131	166	CF$_3$CH=CCl$_2$	164
129	131	79	81	50	CF$_2$Br$_2$	208
130	95	132	75	50	Chlorofluorobenzene	130
131	31	28	44	69	CF$_2$ClCFClCF$_2$C(O)OH	161
131	31	69	93	181	CF$_3$CF$_2$CF=CF$_2$	200
131	31	69	147	93	CF2=CFCF$_2$Cl	166
131	69	100	31	181	Perfluorocyclohexane	281
131	69	100	150	31	CF$_3$CF=CF$_2$	150
131	69	147	101	93	CF$_2$=CFCF$_2$CFCl$_2$	232
131	69	181	31	93	CF$_3$(CF$_2$)$_4$CF=CF$_2$	350
131	69	181	200	31	CF$_3$CF=CFCF$_3$	200
131	75	69	225	175	![structure] FC=CH, F$_2$C, CF$_2$, CF$_2$-CF$_2$	244
131	100	31	69	181	Perfluorocyclopentane	231
131	100	69	31	159	Perfluorocyclohexene oxide	231
131	113	69	31	100	Nonafluorocyclopentane	213
131	133	117	119	95	CCl$_3$CH$_2$Cl	131
131	147	69	31	149	Chloroperfluorocyclohexane	247
132	134	82	84	47	CF$_2$=CCl$_2$	132
133	117	119	135	51	CHF$_2$CCl$_3$	168
135	85	137	129	131	CF$_2$ClCF$_2$Br	214
135	101	85	103	69	CF$_3$CFCl$_2$	151
143	193	69	93	124	CF$_3$C≡CCF$_2$CF$_3$	193
145	126	96	144	51	Trifluoromethylbenzene	146
145	147	85	—	—	CF$_2$ClCH=CCl$_2$	180
147	69	149	31	182	CF$_3$CCl=CClF	182
147	145	31	149	79	CFClBr$_2$	189
147	216	149	69	197	CF$_3$CCl=CFCF$_3$	216
148	150	113	115	47	CFCl=CCl$_2$	148
151	51	69	129	131	CF$_2$BrCF$_2$CHF$_2$	230
151	117	153	119	101	CF$_3$CCl$_3$	167

(*continued*)

Table 14.4. *(continued)*

Base peak	Four next most abundant peaks				Compound	Highest m/z peak >1%*
151	132	69	101	201	CF$_3$CCl$_2$CF$_3$	220
155	205	224	124	69	Perfluoro-1,4-cyclohexadiene	224
161	142	111	114	162	CF$_3$C$_6$H$_4$NH$_2$	161
162	93	69	243	143	Perfluorocyclohexene	262
166	164	129	131	168	CCl$_2$=CCl$_2$	164
171	152	121	170	75	CF$_3$C$_6$H$_4$CN	171
173	145	189	95	75	CF$_3$C$_6$H$_4$C(O)NH$_2$	189
175	177	95	51	112	CF$_2$BrCHFCH$_2$Br	254
179	181	31	129	131	CF$_2$BrCF$_2$Br	258
179	181	85	183	216	CF$_2$ClCCl=CCl$_2$	214
185	85	131	129	69	CF$_3$CFBrCF$_2$Cl	229
185	117	183	119	101	CCl$_3$CFCl$_2$	183
186	117	31	93	155	Hexafluorobenzene	186
189	77	120	92	65	CF$_3$C(O)NHC$_6$H$_5$	189
194	109	196	69	85	1,2-Dichlorooctafluoro-1-hexene	294
196	69	127	177	31	CF$_3$I	196
197	195	147	145	31	CF$_2$BrCFClBr	274
202	55	116	31	28	Chloropentafluorobenzene	202
208	308	131	31	93	C$_4$F$_7$I	308
214	195	145	164	75	C$_6$H$_4$(CF$_3$)$_2$	214
227	254	127	100	31	CF$_2$ICF$_2$I	254
229	231	129	131	69	CF$_3$CFBrCF$_2$Br	229
234	69	133	64	165	CF$_3$SSSCF$_3$	234
263	282	213	225	163	C$_6$H$_3$(CF$_3$)$_3$	282

*Most abundant isotopes.

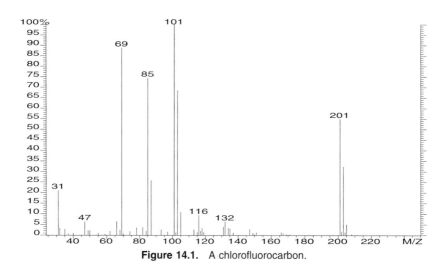

Figure 14.1. A chlorofluorocarbon.

the highest mass observed (greater than 1%). An intense ion at m/z 45 is further confirmation of the acid.

G. Perfluorinated Ketones

Molecular ion is usually not observed with perfluorinated ketones, but may be deduced by adding 19 Daltons to the highest mass observed in the case of perfluoroacetone and 69 Daltons in the case of perfluorodiethyl ketone. A characteristic fragment ion results from α-cleavage:

Example 14.1: $CF_3C(O)$—m/z 97
          $CF_3CF_2C(O)$—m/z 147

m/z 28 (CO) is usually detectable after subtracting the background.

H. Chlorofluoroacetones

1. General formula: $R_1C(O)R_2$

2. Molecular ion: The molecular ions of chlorofluoroacetones usually are not observed, but can be deduced by adding 35 Daltons to the highest-mass peak observed. Cleavage on either side of the carbonyl group defines R_1 and R_2. By deducing the molecular weight and looking for the R_1 and R_2 groups, the particular chlorofluoroacetone can be identified. The chlorofluoroacetones and their molecular weights are given in the following list:

Compound	MW
$CF_3C(O)CF_2Cl$	182
$CF_3C(O)CFCl_2$	198
$CF_2ClC(O)CF_2Cl$	198
$CF_3C(O)CCl_3$	214
$CF_2ClC(O)CFCl_2$	214
$CFCl_2C(O)CFCl_2$	230
$CF_2ClC(O)CCl_3$	230
$CFCl_2C(O)CCl_3$	246

Chapter 15

Gases

I. GC Separations

A. Capillary Columns

1. Oxygen, nitrogen, methane, and carbon monoxide
   25 m Molecular Sieve 5A Plot column at room temperature.
   (Note: This column separates oxygen and nitrogen better than
   the Molecular Sieve 13X column. Neither column separates
   argon and oxygen.)

2. Nitrous oxide, carbon dioxide, and nitric oxide
   25 m GS-Q column or 25 m Poraplot Q column at room temperature.

3. Air, carbon monoxide, methane, and carbon dioxide
   25 m Carboplot OO7 column at 60°.

4. Carbon dioxide, carbonyl sulfide, hydrogen cyanide, propylene,
   and butadiene
   25 m GS-Q column or 25 m Poraplot Q column, 60–200° at
   6°/min.

5. Methane, carbon dioxide, ethylene, propylene, and propane
   25 m Poraplot R column, 30–100° at 5°/min.

6. Methane, ethane, ethylene, propane, propylene, acetylene, iso-butane, and *n*-butane
   25 m Poraplot Alumina column, 35–140° at 6°/min.

7. Air, hydrogen sulfide, carbonyl sulfide, sulfur dioxide, methyl mercaptan, and carbon disulfide
   25 m GS-Q column or 25 m Poraplot Q column, 60–200° at 8°/min.

8. CHF_3, CH_2F_2, CH_3CF_3, CHF_2Cl, CF_2Cl_2, CH_2FCl, CHF_2CH_2Cl, and $CHFCl_2$
   25 m Poraplot Q column, 60–150° at 5°/min.

B. Packed Columns

1. Tetrafluoromethane, hexafluoroethane, tetrafluoroethylene, difluoropropane, hexafluorocyclopropane, and *n*-hexafluoro-propylene
   2 m silica gel column, 35–180° at 10°/min.

2. Carbon monoxide and carbon dioxide
   2 m silica gel column, 60–200° at 5°/min.

3. Carbon dioxide, sulfur dioxide, carbonyl sulfide, acetylene, and hydrogen sulfide
   2 m silica gel column at room temperature.

II. General Information

A gas analysis usually involves some or all of the common gases: O_2, N_2, CO, CO_2, $C_1–C_5$ hydrocarbons, low-boiling fluorinated compounds, sulfur compounds, and so forth. Separating or passing the effluent from the GC column through a sensitive thermal conductivity detector (TCP) before entering the MS enables the qualitative and quantitative analyses of unknown gas mixtures from ppm levels to percentage levels. If elaborate column switching systems are not available, two GC/MS runs may be required on two different GC columns. For instance, CO_2 and the $C_2–C_5$ hydrocarbons are adsorbed on the Molecular Sieve 5A column while separating H_2, O_2, N_2, CH_4, and CO. A second GC/MS run is performed using GS-Q or Poraplot Q, which separates CO_2 and the $C_2–C_5$ hydrocarbons from composite peaks of H_2, O_2, N_2, CH_4, and CO. By performing two GC/MS runs on two different columns, a complete gas analysis can be achieved. Remember that if a sufficient number of analyses are made, the Molecular Sieve 5A column will have to be replaced.

Chapter 16

Glycols

I. GC Separations

A. Underivatized Glycols

 1. Capillary columns

   a. Propylene glycol, ethylene glycol, dipropylene glycol, diethylene glycol, triethylene glycol, and tetraethylene glycol
   30 m NUKOL column, 60–220° at 10°/min.

   b. Ethylene glycol, diethylene glycol, triethylene glycol, and tetraethylene glycol
   30 m DB-5 column, 100 (2 min)–200° at 10°/min.

   c. 2,3-butanediol, 1,2-butanediol, 1,3-butanediol, and 1,4-butanediol
   30 m DB-Wax column, 50 (5 min)–150° at 15°/min.

   d. Isobutylene glycol, propylene glycol, dipropylene glycol, and tripropylene glycol
   30 m DB-Wax column, 180–195° at 2°/min.

 2. Packed columns

   a. Water, ethylene glycol, 1,2-propanediol, 1,3-propanediol, 1,3-butanediol, 1,4-butanediol, and glycerol
   2 m Porapak Q column at 220°. Run for 30 min.

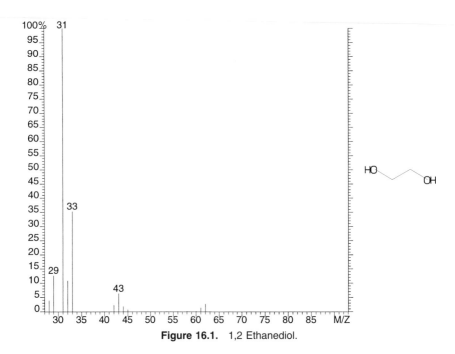

Figure 16.1. 1,2 Ethanediol.

b. Ethylene glycol impurities: methanol, acetaldehyde, dime-thoxyethane, and 2-methyl-1,3-dioxolane
2 m Porapak Q column at 150°.

c. Ethylene glycol, 1,3-propanediol, 1,4-butanediol, 1,5-pen-tanediol, 1,6-hexanediol, 1,7-heptanediol, 1,8-octanediol, 1,9-nonanediol, and 1,10-decanediol
2 m Tenax-GC column, 100–300° at 6°/min.

d. Butyl alcohol, ethylene glycol monoethyl ether, cyclohexanol, benzyl alcohol, butyl carbitol, and dibutyl phthalate
2 m Tenax-GC column, 100–300° at 10°/min.

II. Derivatization of Dry Glycols and Glycol Ethers

A. TMS Derivatives

Add 0.25 ml of Tri-Sil Z reagent to 1–2 mg of sample. Stopper and heat the mixture at 60° for 5–10 min.

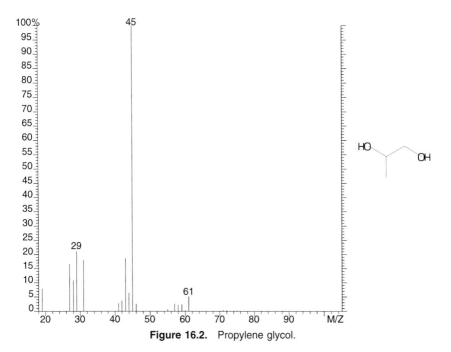

Figure 16.2. Propylene glycol.

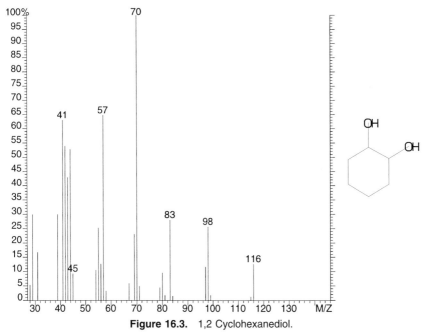

Figure 16.3. 1,2 Cyclohexanediol.

B. GC Separation of TMS Derivatives of Glycols and Glycol Ethers*

1. Ethylene glycol-TMS, diethylene glycol-TMS, and glycerol-TMS
    30 m DB-1 column or 30 m DB-5 column, 60–175° at 4°/min.

III. Mass Spectral Interpretation

A. Glycols and Glycol Ethers (Underivatized)

In the mass spectra of di- and tri-ethylene glycol, the largest peak observed is at m/z 45. Because oxygen is present, an m/z 31 ion is also expected.

B. TMS Derivatized Glycols

The best way to identify the MWs of unknown glycols and glycol ethers is to examine the mass spectra of the TMS derivatives. The TMS derivatives are identified in the chromatogram by plotting masses 73 and 147. The higher of the two high-mass peaks, which are 15 Daltons apart, is the molecular ion of the TMS derivative. If no peaks are separated by 15 Daltons, add 15 to the highest mass peak observed to deduce the MW.

C. Sample Mass Spectra

Figure 16.1 shows the presence of both m/z 31 and 45, indicating that oxygen is present in the molecule. The ion at m/z 33 is characteristic of some hydroxy compounds and has been suggested by McLafferty[1] to occur by β-bond fragmentation accompanied by the rearrangement of two hydrogen atoms. The ion at m/z 62 could be the molecular ion of ethylene glycol. This can be confirmed by preparing the TMS derivative, which will show the presence of two hydroxy functions and confirm the MW assignment.

Figure 16.2 is the mass spectrum of propylene glycol and shows the presence of an abundant m/z 45 ion. A library search will provide strong evidence that this compound is propylene glycol. Preparation of a TMS derivative will confirm this assignment.

Figure 16.3 is the mass spectrum of 1,2-cyclohexanediol and shows a strong molecular ion at m/z 116. The ions at m/zs 31 and 45 suggest oxygen, and the strong fragment ion at m/z 98 (M-18) is characteristic of many diols.

Reference

1. McLafferty, F. W. *Mass Spectrometry of Organic Ions.* New York: Academic Press, 1963.

*The TMS derivatives cannot be injected into a DB-Wax column.

Chapter 17

Halogenated Compounds (Other Than Fluorinated)

I. GC Separations

A. Saturated and Unsaturated Halogenated Compounds

1. Capillary columns

   a. Methyl chloride, vinyl chloride, methyl bromide, ethyl chloride, dichlorethanes, chloroform, carbon tetrachloride, bromochloromethane, and similar compounds
   30 m DB-624 column 35 (5 min)–140° at 5°/min or 25 m CP-Sil 13CB column for halocarbons, 35 (5 min)–80° at 2°/min.

   b. Most dichlorobutenes and trichlorobutenes
   30 m DB-Wax column, 70–200° at 4°/min.

2. Packed columns

   a. Many halogenated compounds from methyl chloride to chlorobenzene
   3 m Carbowax 1500 column on Carbopack C, 60 (4 min)–170° at 8°/min.

   b. Methyl chloride, F-22, methyl bromide, F-12, F-21, methyl iodide, methylene chloride, F-114, F-11, chloroform, F-113, 1,2-dichloroethane, 1,1,1-trichloroethane, and so forth
   3 m SP-1000 column on Carbopack B, 40–175° at 4°/min.

145

    c. Vinyl chloride in air
1 m 60–80 mesh Alumina at 110° or 3 m 10% Carbowax 1540 column on 80–100 mesh Chromosorb W at 50°.

B. Halogenated Aromatics

  1. Capillary columns

    a. Chlorobenzenes

      1. 50 m DB-Wax column, 100 (8 min)–200° at 8°/min.

      2. *o*-, *m*-, and *p*-Isomers
50 m DB-Wax column at 125°.

      3. 1,3-Dichlorobenzene, 1,4-dichlorobenzene, 1,2-dichlorobenzene, 1,2,4-trichlorobenzene, and hexachlorobenzene
30 m DB-1301 column, 75–210° at 10°/min.

    b. Chlorotoluenes

      1. Isomers of chloromethylbenzenes
30 m CP-Sil 88 column, 50 (6 min)–200° at 4°/min.

    c. Chlorinated biphenyls

1. **Molecular Formula***	**Accurate Mass Value**
$C_{12}H_9Cl$	188.0393
$C_{12}H_8Cl_2$	222.0003
$C_{12}H_7Cl_3$	255.9613
$C_{12}H_6Cl_4$	289.9224
$C_{12}H_5Cl_5$	323.8834
$C_{12}H_4Cl_6$	357.8444
$C_{12}H_3Cl_7$	391.8054
$C_{12}H_2Cl_8$	425.7665
$C_{12}H_1Cl_9$	459.7275
$C_{12}Cl_{10}$	493.6885

30 m DB-17 column, 150 (4 min)–260° at 4°/min. or 30-m DB-1 column, 60 (4 min)–260° at 20°/min.

      2. Arochlors 1016, 1232, 1248, and 1260. Arochlors have been analyzed under these GC conditions at the 50 ppm level

*Not all of these isomers are completely separated by the conditions given here, but all are readily detected by plotting the masses of the molecular ions. Pesticides can interfere with polychlorinated biphenyl (PCB) analysis.

using electron impact ionization (EI). At lower concentrations of PCBs, negative chemical ionization (CI) should be considered 50 m CP-Sil-88 column, 150 (4 min)–225° at 4°/min.

d. Chloronitrotoluenes and dichloronitrotoluenes
   30 m DB-17 column, 75–250° at 6°/min.

e. Halogenated toluene diisocyanates: chloro-, bromo-, dichloro-, and trichlorotoluene diisocyanates
   30 m DB-1 column, 70–225° at 4°/min.

f. Haloethers: bis(2-chloroethyl) ether, bis(2-chloroisopropyl) ether, and bis(2-chloroethoxy) methane
   30 m DB-1301 column, 100 (5 min)–130° at 10°/min, 130–250° at 15°/min.

g. Halogenated pesticides: lindane, heptachlor, aldrin, chlordane, dieldrin, DDT, and similar compounds (see Chapter 25)
   30 m DB-5 column, 60–300° at 4°/min.

2. Packed columns

a. Chloronaphthalenes: 1-, and 2-chloronaphthalenes
   3 m packed SP-1200 column + Bentone-34, 60–170° at 6°/min.

II. Mass Spectra of Halogenated Compounds (Other Than Fluorinated Compounds)

A. Aliphatic Halogenated Compounds

The presence of chlorine and/or bromine is easily detected by their characteristic isotopic patterns (see Appendix 11). As in many aliphatic compounds, the abundance of the molecular ion decreases as the size of the R group increases. For example, in the EI mass spectra of methyl chloride and ethyl chloride, the molecular ion intensities are high, whereas in compounds with larger R groups such as butyl chloride, the molecular ion peak is relatively small or nonexistent.

The highest mass peaks observed in the mass spectra of alkyl chlorides may correspond to the loss of HX or X (loss of HI is seldom observed), depending on the structure of the molecule. In order to deduce the molecular ion, add the mass of X or HX to the mass at which the highest mass peak is readily observed. (Note that higher mass ions having the isotope pattern of X may be present

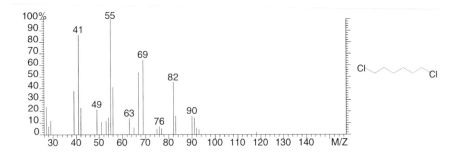

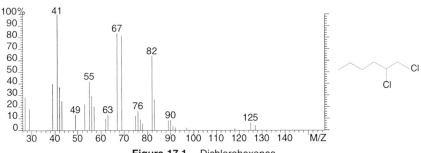

Figure 17.1.   Dichlorohexanes.

in low abundances.) Try to select the structural type by examining characteristic low-mass fragment ions. For example, if one observes a large m/z-49 peak with a m/z-51 peak that is approximately one-third its intensity, this may indicate that a terminal chlorine is present in an alkyl halide. An intense m/z 91 peak (with the isotope pattern for a chlorine) represents a C_6–C_{18} 1-alkyl chloride. An intense m/z 135 peak (with the isotope pattern for bromine) suggests a C_6–C_{18} 1-alkyl bromide.

B. Sample Mass Spectrum of an Aliphatic Halogenated Compound

1-Alkyl halides are more easily identified than the dichloro halides. Figure 17.1 shows the mass spectra of two dichlorohexanes. Long-chain saturated halides may also lose an alkyl portion from the molecular ion, such as 15, 29, 43, 57, 71, 85, and 99 Daltons. These can be identified as halogenated compounds, but it is difficult to deduce their molecular weights without CI or negative CI.

C. Aromatic Halogenated Compounds

The molecular ion peaks in the mass spectra of aromatic halogenated compounds are fairly intense. The molecular ion abundances

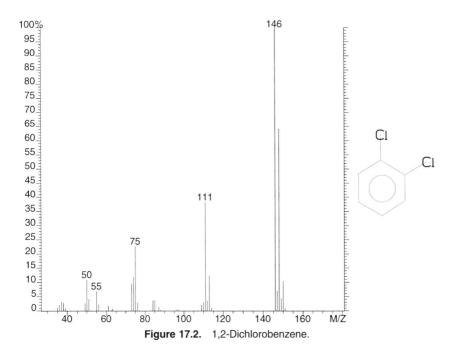

Figure 17.2. 1,2-Dichlorobenzene.

decrease with increasing length of the side chain. If an intense ion at *m/z* 91 (tropylium ion) is present, then the halogen is on the alkyl side chain rather than the ring. If the halogen is on the ring, the loss of the halogen from the molecular ion will result in a small peak. The loss of HX is favored when a CH_3 group is ortho to the halogen. The loss of $-CH_2X$ is observed if the group on the ring is $-CH_2CH_2X$.

D. Sample Mass Spectrum of an Aromatic Halogenated Compound

Figure 17.2 is an example of a mass spectrum of an aromatic dichloro compound. The intensity of the molecular ion indicates that an aromatic compound is present. The isotope pattern is that of two chlorines, and subtracting 70 mass units from the molecular ion gives the formula C_6H_4. (See Example 2.3 in Chapter 2 for another example of isotope abundances in the molecular ion region.)

Chapter 18

Hydrocarbons

I. GC Separation of Hydrocarbons

A. Saturated and Unsaturated Aliphatic Hydrocarbon Compounds

1. C_1–C_5 hydrocarbon compounds

    a. 50 m Al_2O_3/KCl Plot column
       40 (1 min)–200° at 10°/min or 60–200° at 3°/min.

    b. 30 to 50 m GS-Q column or Poraplot Q column
       35–100° at 10°/min. 100° (5 min); 100–200° at 10°/min, or 40
       (2 min)–115° at 10°/min.

2. C_4 hydrocarbon isomers
   2 m Picric acid column on Carbopak C, 30–100° at 2°/min.

3. C_4–C_{12} hydrocarbon compounds
   100 m SQUALANE column, 40–70° at 0.5°/min.

4. Natural gas
   30 m GS-Q column at 75°.

5. C_5–C_{100} hydrocarbon compounds
   30 m SIM DIST-CB Chrompack column, 100–325° at 10°/min;
   injection port at 325°.

B. Low-Boiling Aromatic Hydrocarbon Compounds

1. Benzene, toluene, ethylbenzene, *p*-xylene, *m*-xylene, *o*-xylene
   30 m CP CW57 CB column, 50–200° at 5°/min.

2. Benzene, toluene, ethylbenzene, *p*-xylene, *m*-xylene, *o*-xylene,
   butylbenzene, styrene, *o*-, *m*-, and *p*-diethylbenzenes
   25 m DB-WAX column, 50 (2 min)–180° at 5°/min.

C. Polynuclear Aromatic Hydrocarbon Compounds

1. Naphthalene, acenaphthane, acenaphthene, fluorene, phenanthrene,* anthracene,* fluoranthene, pyrene, benzanthracene,
   benzophenanthrene, and benzpyrene
   30 m DB-1 column, 120 (6 min)–275° at 10°/min.

2. Phenanthrene, anthracene, and other polynuclear aromatic compounds.
   30 m DB-1301 column, 100–260° at 5°/min.

D. Gasoline

Gasoline contains more than 250 components of a mixture of
C_4–C_{12} hydrocarbons, which varies in concentration from batch
to batch. Some of these components are: *iso*butane, *n*-butane,
*iso*pentane, *n*-pentane, 2,3-dimethylbutane, 3-methylpentane, *n*-
hexane, 2,4-dimethylpentane, benzene, 2-methylhexane, 3-meth-
ylhexane, 2,2,4-trimethylpentane, 2,3,4-trimethylpentane, 2,5-
dimethylhexane, 2,4-dimethylhexane, toluene, 2,3-dimethylhexane,
ethylbenzene, methylethylbenzenes, *m*-, *p*-, and *o*-xylene, trimeth-
ylbenzenes, naphthalene, methylnaphthalenes, and dimethylnaph-
thalenes

1. 30 m DB-1701 column, 35 (10 min)–180° at 6°/min.

2. 30 m DB-1 column, 35 (10 min)–200° at 6°/min.

II. Mass Spectra of Hydrocarbon Compounds

A. *n*-Alkanes

The molecular ions decrease in intensity with increasing chain
length but are still detectable at C_{40}. In contrast to branched alkanes,
the loss of a methyl group is not favored for *n*-alkanes. Usually the

*Not separated using the DB-1 column.

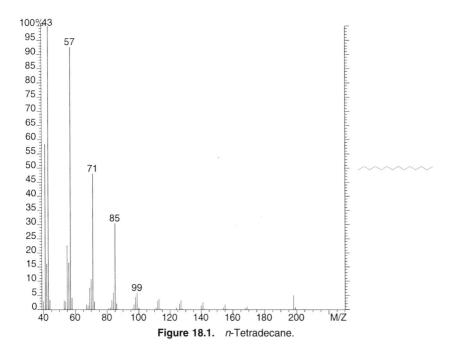

Figure 18.1. *n*-Tetradecane.

first fragment ion below the molecular ion is at mass M − 29. Compounds of C_4 and higher show a base peak at *m/z* 43 or 57. Alkanes yield a series of peaks differing by 14 Daltons (e.g., 43, 57, 71, 85, etc.).

B. Sample Mass Spectrum of an Alkane

Figure 18.1 is an example of a C_{14} aliphatic hydrocarbon. The molecular ion is observed along with the M − 29 fragment and the low-mass ions 43, 57, and so forth, separated by 14 Daltons.

C. Branched Alkanes

The molecular ion intensity decreases with increased branching, therefore the molecular ion peak may be nonexistent. The loss of 15 Daltons from the molecular ion indicates a methyl side chain. The mass spectra of branched alkanes are dominated by the tendency for fragmentation at the branch points, and hence are difficult to interpret.

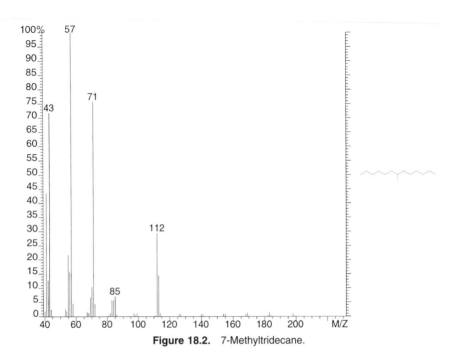

Figure 18.2. 7-Methyltridecane.

D. Sample Mass Spectrum of a Branched Alkane

Figure 18.2 is the mass spectrum of a branched hydrocarbon. Note the intensity of the molecular ion peak and the presence of an M − 15 peak. The M − 15 peak is typical, particularly if the side chain is a methyl group. The position of the side chain is indicated by the m/z 112 and 113 ions ($CH_3(CH_2)_5CHCH_3$). Usually the site of branching is more difficult to establish.

E. Cycloalkanes

The molecular ion intensity is more abundant in cycloalkanes than in straight-chain alkanes containing the same number of carbon atoms. Fragmentation of the ring is usually characterized by the loss of 28 and 29 mass units, respectively. The tendency to lose even-numbered fragment ions, such as C_2H_4, produces mass spectra that contain a greater number of even-numbered mass ions than the mass spectra of straight- or branched-chain hydrocarbons.

A saturated ring with an aliphatic side chain favors cleavage at the bond connecting the ring to the rest of the molecule. Compounds containing cyclohexyl rings fragment at m/z 83, 82, and 81, corresponding to ring fragmentation and the loss of one and two hydrogen atoms.

F. Alkenes

Molecular ions are usually intense for low-molecular weight compounds. Alkyl cleavage with the charge remaining on the unsaturated portion is very often the base peak. A series of fragment ions with *m/z* 41, 55, 69, 83, and so forth are characteristic. Methods are available to locate the position of the double bond in aliphatic compounds.[1]

G. Alkylbenzenes

Alkylbenzenes have molecular ions at the following *m/z* values: 92, 106, 120, 148, and so forth. The molecular ion intensity decreases with increasing alkyl chain length, but can be detected up to at least C_{16}. Characteristic fragment ions are: *m/z* 39, 50, 51, 52, 63, 65, 76, 77, and 91.

m/z 91

An excellent method for identifying alkylbenzenes was developed by Meyerson[2] and should be consulted.

H. Polynuclear Aromatic Hydrocarbons

Unsubstituted polynuclear aromatic hydrocarbons show intense molecular ions. The aklylated polynuclear aromatics and the alkylated benzenes fragment similarly:

m/z 141

Characteristic fragment ions of alkylnaphthalenes are *m/z* 141 and 115. Using the GC conditions mentioned previously, most of the molecular ions of the following compounds can be found by plotting the accompanying accurate mass values (listed in order of elution):

Compounds	Accurate Mass Value
Naphthalene	128.0626
Acenaphthene	154.0783
Fluorene	166.0783
Phenanthrene	178.0783
Anthracene	178.0783
Fluoranthene	202.0783
Pyrene	202.0783
Benzanthracene	228.0939
Chrysene	228.0939
Benzpyrene	252.0939

If all isomers are present, identification is straightforward; however, if only one isomer is present, standards may have to be injected into the GC/MS to obtain retention times under the GC conditions used.

References

1. Schneider, B., and Budzikiewicz, H. A facile method for the localization of a double bond in aliphatic compounds. *Rapid Commun. Mass Spectrom.*, *4*, 550, 1990.
2. Meyerson, S. Correlations of alkylbenzene structures with mass spectra. *Applied Spectros.*, *9*, 120, 1955.

Chapter 19

Isocyanates

I. GC Separations

A. Toluene diisocyates (TDI), xylene isocyanates, chloro-TDI, bromo-TDI, dichloro-TDI, and trichloro-TDI
   30 m DB-1 column, 70–225° at 4°/min.

B. *m*-Phenylene-diisocyanate, toluenediisocyanate, xylenediisocyanate, butylated hydroxytoluene, 5-chlorotoluenediisocyanate, and methylene-bis-(4-cyclohexylisocyanate)
   30 m DB-1 column, 100–300° at 8°/min.

C. Xylene diisocyanates
   30 m DB-1 column, 70–235° at 4°/min.

II. Mass Spectral Interpretation

A. Mass Spectra of Aliphatic Isocyanates

   1. General formula: RNCO

   2. Molecular ion: The molecular ions of aliphatic isocyanates are observed up to C_8.

   3. Fragmentation: Characteristic fragments of aliphatic isocyanates include: m/z 56 (CH_2NCO), 70 (CH_2CH_2NCO), 84 ($CH_2CH_2CH_2NCO$), and so forth.

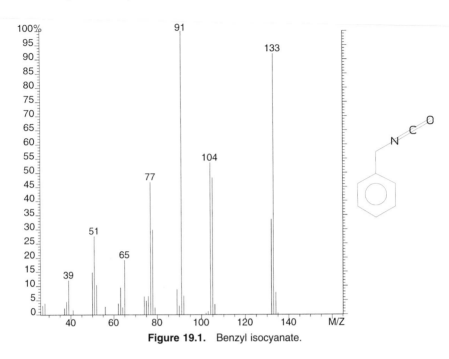

Figure 19.1. Benzyl isocyanate.

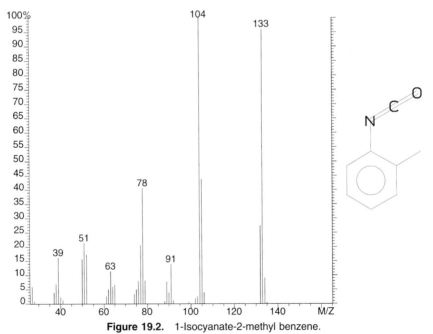

Figure 19.2. 1-Isocyanate-2-methyl benzene.

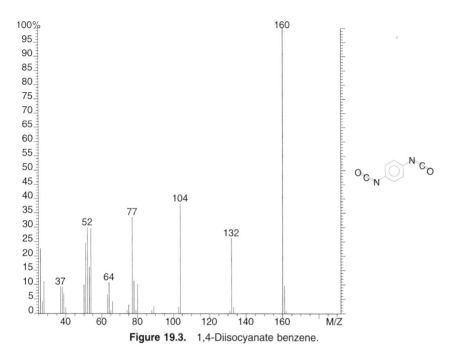

Figure 19.3. 1,4-Diisocyanate benzene.

B. Mass Spectra of Aromatic Isocyanates

  1. General formula: ArNCO

  2. Molecular ion: The molecular ions of the aromatic isocyanates and diisocyanates are usually always observed, depending on the length of the alkyl groups on the ring.

  3. Fragmentation: Losses from the molecular ion include: m/z 28 (CO), 29 (H + CO), and 55 (CO + HCN). Sometimes loss of hydrogen is observed, particularly if there is a methyl group on the ring.

C. Sample Mass Spectra

  Figures 19.1 and 19.2 are the mass spectra for the aromatic isocyanates. Losses of 28 (CO) and 29 (H + CO) Daltons are observed. The intensity of m/z 91 (M-NCO) ion favors the benzyl structure in Figure 19.1.

  Figure 19.3 is an aromatic diisocyanate. The observed successive losses of 28 and 56 Daltons are similar to the losses found with quinone or anthraquinone.

Chapter 20

Ketones

I. GC Separation of Ketones

A. Capillary Columns

    1. Most ketones from acetone to 3-octanone
       50 m DB-5 column, 40 (3 min)–250° at 10°/min (nonselective).

    2. TMS derivatives of multifunctional ketones
       30 m DB-210 column (selective for ketones), 40–220° at 6°/min.

    3. Cyclohexanone, cyclohexanol, cyclohexenone, dicyclohexyl ether, cyclohexyl valerate, cyclohexyl caproate, valeric acid, and caproic acid
       30 m FFAP column, 60–200° at 6°/min.

B. Packed Columns

    1. Acetone, 2-butanone, 3-methyl-2-butanone, 2-pentanone, 3,3-dimethyl-2-butanone, cyclopentanone, 3-heptanone, 4-methyl-cyclohexanone, 2-octanone, and acetophenone
       2 m Porapak Q column, 100–200° at 10°/min.

II. Derivatives of Ketones

A. Methoxime Derivatives

This derivative is useful for determining the presence and number of keto groups as well as for protecting the ketone from enolization. Some diketones that polymerize readily, such as 2,3-butanedione, should be freshly distilled and the methoxime derivatives should be prepared.

1. Preparation of methoxime derivatives: Add 0.5 ml of MOX reagent to the sample. Heat at 60° for 3 hours. Evaporate the reaction mixture to dryness with clean, dry nitrogen. Dissolve in the minimum amount of ethyl acetate. Some solids will not dissolve.

B. Other Derivatives

The ketone group can be reduced to an alcohol that can then be silylated. This procedure has been used to identify the keto group in carbohydrates.

1. Reduction and preparation of derivatives: Concentrate the aqueous mixture to 0.5 ml. Add 20 mg of sodium borohydride dissolved in 0.5 ml of ion exchange water. Let this solution stand at room temperature for 1 hour. Destroy the excess sodium borohydride by adding acetic acid until gas evolution stops. Evaporate the solution to dryness. Add 5 ml of methanol and evaporate again to dryness.
Prepare the TMS derivative by adding 250 μl of MSTFA reagent and then heat at 60° for 5 min.

III. Mass Spectra of Ketones

A. Aliphatic Ketones

1. General formula: $RC(O)R'$

2. Molecular ion: The MW of aliphatic ketones can be determined from its prominent molecular ion. In general, the intensity of the molecular ions of ketones is greater for C_3–C_8 than for C_9–C_{11}. A molecular ion is usually observed for methoxime derivatives.

3. Fragmentation: The unsymmetrical ketones usually yield four major fragment ions from cleavage on either side of the carbonyl group: R^+, $RC\equiv O^+$, R'^+, and $R'C\equiv O^+$. The oxygen-containing fragment ions are usually more intense than the corresponding R^+ and R'^+ ions. Loss of 31 Daltons from the molecular ion of methoxime derivatives is observed.

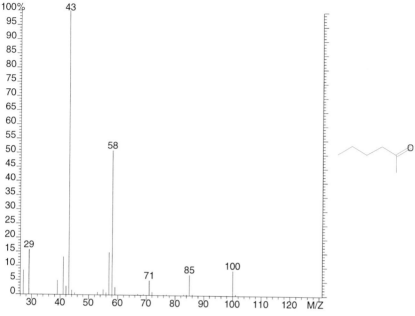

Figure 20.1. 2-Hexanone.

4. Characteristic fragment ions: Aliphatic ketones also give abundant McLafferty rearrangement ions at m/z 58, 72, 86, and so forth. Methyl ketones produce an abundant ion at m/z 43. Low-intensity ions at m/z 31, 45, 59, 73, and so on reveal oxygen in the unknown ketone and are especially useful in distinguishing ketone spectra from isomeric paraffin spectra. Subtract 43 from the mass of the rearrangement ion to determine R.

B. Sample Mass Spectrum of Aliphatic Ketones

In the mass spectrum of 2-hexanone (Figure 20.1), the molecular ion is apparent at m/z 100, which can be confirmed by preparing the methoxime derivative. The compound type is verified by the presence of m/z 43 and 58.

C. Cyclic Ketones

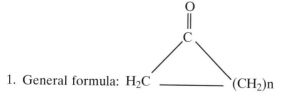

1. General formula:

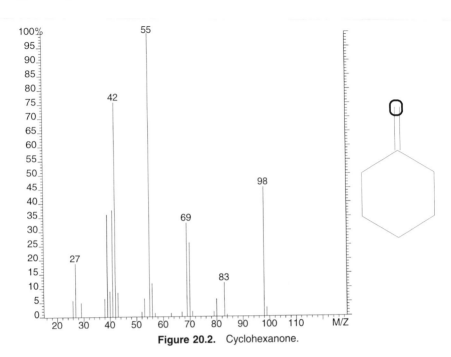

Figure 20.2. Cyclohexanone.

2. Molecular ion: Molecular ions of cyclic ketones are relatively intense. Characteristic fragment ions of cyclic ketones occur at m/z 28, 29, 41, and 55. Cyclic ketones also lose CO and/or C_2H_4 (m/z 28) from the molecular ion (C_6 and higher). Low-abundance ions corresponding to loss of H_2O are frequently observed. Keto-steroids are a special class of cyclic ketones and have abundant molecular ions.

D. Sample Mass Spectrum of Cyclic Ketones

Figure 20.2 represents a cyclic ketone as shown by the loss of 28 and 29 Daltons from the molecular ion, which is characteristic of cyclic ketones. Also, the loss of 18 Daltons from the molecular ion frequently is observed.

E. Aromatic Ketones

$$\overset{\text{O}}{\underset{\|}{}}$$

1. General formula: $ArC - R$

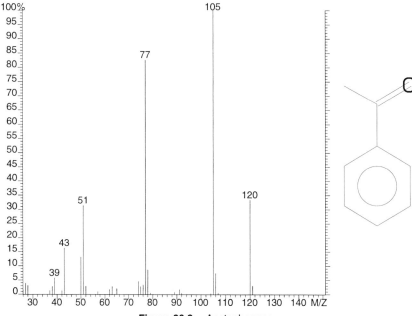

Figure 20.3. Acetophenone.

2. Molecular ion: The molecular ion is always present.

3. Fragmentation: Fragmentation occurs on both sides of the carbonyl group. For example, in acetophenone, the major ions occur
   at masses 77, 105, and 120 (see Figure 20.3). Ions at m/z 39, 50,
   and 51 also suggest the presence of an aromatic ring. Aromatic
   compounds, such as quinone, tetralone, and anthraquinone,
   readily lose CO.

Chapter 21

Nitriles

I. GC Separation of Nitriles

A. General

   1. Capillary columns

      a. Ethyl succinonitrile, 1,1-cyanophenylethane, 1,2-cyanophenylethane, methylglutaronitrile, and adiponitrile
        30 m DB-210 column, 50–225° at 10°/min.

      b. Dicyanobutenes
        30 m DB-210 column, 50–225° at 6°/min.

      c. Isoxazole, fumaronitrile, propanedinitrile, and ethylene nitrile
        30 m DB-Wax column, 100–225° at 8°/min.

   2. Packed columns

      a. Ammonia, acetonitrile, *n*-methyethyleneimine, and propionitrile
        2 m Chromosorb 103 column, 50 (5 min)–150° at 8°/min.

167

B. Low-Boiling Nitriles

1. Acetonitrile, acrolein, acrylonitrile, and propionitrile
30 m Gas Q column or 30 m Poraplot Q column, 40 (2 min)–115° at 10°/min.

2. Acetonitrile, acrylonitrile, methacrylonitrile, isobutyronitrile, and butyronitrile
30 m Gas Q column or 30 m Poraplot Q column, 40 (2 min)–115° at 10°/min.

II. Mass Spectra

A. Aliphatic Mononitriles

1. General formula: RCN

2. Molecular ion: The aliphatic mononitriles may not show molecular ions when R is greater than C_2.

3. Fragmentation: The saturated aliphatic mononitriles with molecular weights greater than 69 are characterized by intense ions at m/z 41, 54, 68, 82, 96, 110, 124, 138, 152, 166, and so forth. Aliphatic nitriles undergo the McLafferty rearrangement producing the m/z 41 ion.

   The aliphatic mononitriles may not show molecular ions, but M − 1, M − 27, or M − 28 are usually observed. Sometimes a loss of 15 Daltons may also be observed. If CH_3 is replaced by CF_3, as in the case of $CF_3CH_2CH_2CN$, a fluorine is first lost from the molecular ion, followed by the loss of HCN (from the M − F ion). This influence of the CF_3 group diminishes as the alkyl chain length increases.

B. Sample Mass Spectrum of an Aliphatic Mononitrile

The mass spectrum in Figure 21.1 exhibits intense ions at m/z 41, 54, 68, and so forth, which suggests an alkyl nitrile. The highest mass ion at m/z 96 is M − 1. Also observed are the ions corresponding to M − 15, M − 27, and/or M − 28.

C. Aliphatic Dinitriles

1. General formula: $NC(CH_2)_xCN$

2. Molecular ion: The molecular ions of aliphatic dinitriles with molecular weights greater than 80 are usually not observed, but

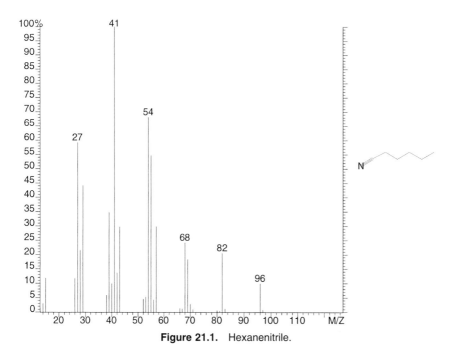

Figure 21.1. Hexanenitrile.

their molecular ions can be deduced by adding 40 (CH_2CN) Daltons to the highest mass of reasonable intensity.

3. Fragmentation: The highest mass ions are usually M − 40 with a less abundant M − 28. Intense peaks at m/zs 41, 54, and 55 also should be present (see Figure 21.2). Dinitriles also may have ions at m/z 82, 96, 110, 124, 138, 152, and so on. Except for adiponitrile and methylglutaronitrile, m/z 68 is of very low intensity in the mass spectra of aliphatic dinitriles. Adiponitrile may be distinguished from methylglutaronitrile by the relative intensities of m/z 41 and 68. If m/z 41 is of greater abundance than m/z 68, the mass spectrum suggests adiponitrile. Conversely, if the abundance of m/z 68 is greater than that of m/z 41, the mass spectrum represents methylglutaronitrile.

D. Aromatic Nitriles and Dinitriles

1. General formula: ArCN and Ar(CN)$_2$

2. Molecular ion: The mass spectra of aromatic nitriles and dinitriles show intense molecular ions (see Figure 21.3).

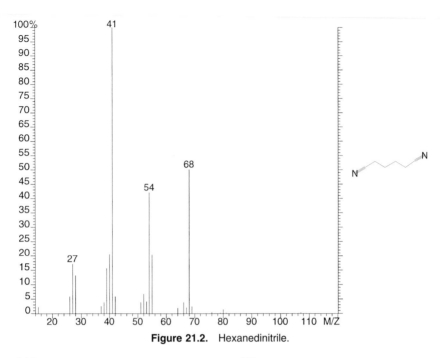

Figure 21.2.   Hexanedinitrile.

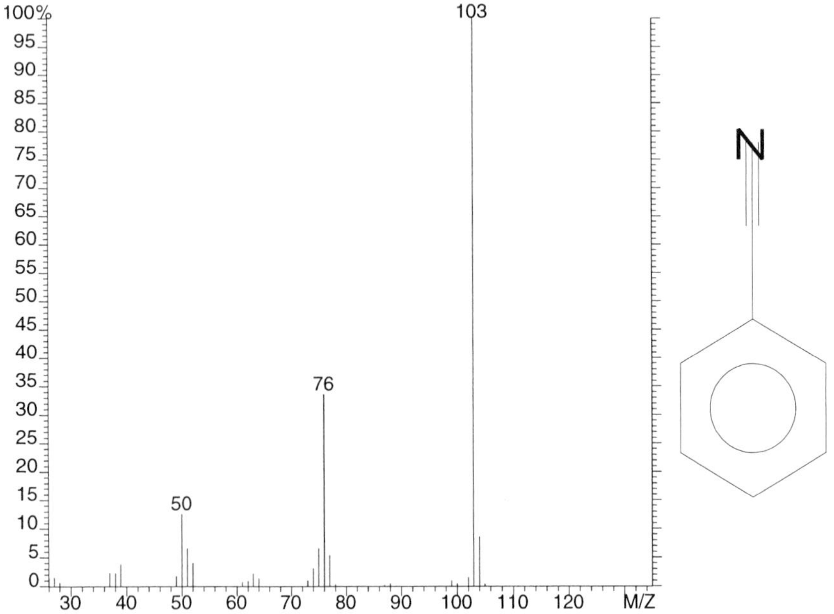

Figure 21.3.   Benzonitrile.

3. Fragmentation: The loss of 27 (HCN) Daltons from the molecular ions of aromatic nitriles and dinitriles is characteristic. The tolunitriles show a loss of hydrogen as well as the loss of HCN and H_2CN. If the methyl group is replaced by a CF_3 group, the loss of 19 (F) and 50 (CF_2) Daltons is very intense, while the loss of 27 Daltons is practically nonexistent. If the nitrile group is on the side chain rather than on the aromatic ring, the loss of 27 (HCN) Daltons occurs.

Chapter 22

Nitroaromatics

I. GC Separation of Nitroaromatics

A. Capillary columns

1. Aniline, nitrobenzene, phenylenediamine isomers, nitroaniline isomers, azobenzene, and azoxybenzene
30 m DB-1 column, 75–250° at 10°/min.

2. Dichlorobenzene, aminotoluene, nitrotoluene isomers, diaminotoluene, and dinitrotoluene isomers
30 m DB-17 column, 100–250° at 8°/min.

3. Nitroanilines
30 m DB-17 column, 75–250° at 10°/min.

4. Nitrophenols, nitroaminotoluenes, *N,N*-dimethylnitroanilines, chloronitrotoluenes, chloronitroanilines, nitronaphthalenes, dinitrotoluenes, dinitroanilines, dinitrophenols, dinitrochlorobenzenes, dichloronitrotoluenes, chlorodinitroanilines, dinitronaphthalenes, and trichloronitrotoluenes
30 m DB-17 column, 75–250° at 10°/min.

B. Dinitrobenzenes
30 m DB-17 column at 225°.

C. Nitrodiphenylamines
30 m DB-225 column or 30 m CPSIL 43CB column, 75–225° at 10°/min.

D. Di(4-nitrophenyl)ether and 1,2-Di(4-nitrophenyl)ethane
30 m DB-17 column, 100–275° at 6°/min.

II. Mass Spectra of Nitroaromatics

A. Nitroaromatics

1. General formula: ArNO

2. Molecular ion: The mass spectra of nitroaromatic compounds are characterized by intense molecular ions.

3. Fragmentation: Some or all of the following fragment ions are observed:
   M − 16 (O), M − 17 (OH for ortho isomers), M − 30 (NO), M − 46 (NO_2), M − 58 (NO + CO), and M − 92 (NO_2 + NO_2).

B. Nitrotoluenes (MW = 137)

The *o*-nitrotoluene isomer is easy to identify because of the loss of OH from the molecular ion. All the nitrotoluenes lose 30 Daltons from their molecular ions. The *m*- and *p*-isomers can be distinguished from each other by the relative abundances of the *m/z* 65 ion versus the molecular ion, particularly if both isomers are present (see Figure 22.1).

C. Nitroanilines (MW = 138)

The mass spectra of all three isomers are different. The ortho isomer loses 17 Daltons (OH, small peak) from its intense molecular ion. The *m*- and *p*-isomers lose 16 Daltons from their molecular ions and can be distinguished by comparing the relative abundances of the *m/z* 65 fragment ion versus their molecular ions. The *m/z* 65 ion is the most abundant ion in the mass spectrum of the *m*-isomer (see Figure 22.2), while the molecular ion at *m/z* 138 is the most abundant ion in the mass spectrum of the *p*-isomer.

D. Nitrophenols (MW = 139)

o-Nitrophenol loses OH from its molecular ion. Both *m*- and *p*-isomers lose 16, 30, and 46 Daltons from their molecular ions. The *m*- and *p*-isomers can be distinguished from each other by the high abundance of the *m/z* 109 ion in the mass spectrum of the *p*-

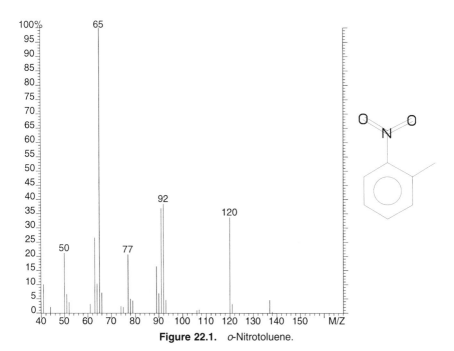

Figure 22.1. *o*-Nitrotoluene.

isomer and its low abundance in the mass spectrum of the *m*-isomer (see Figure 22.3).

E. Nitroaminotoluenes (MW = 152)

Nitroaminotoluenes have abundant molecular ions and a characteristic *m/z* 77 fragment ion. Other intense ions that are frequently observed are *m/z* 135, 107, 106, 104, 79, and 30. The loss of 17 Daltons is especially abundant when the methyl and nitro groups are ortho to each other.

F. *N,N*-dimethylnitroanilines (MW = 166)

Again, the *o*-isomer of *N,N*-dimethylnitroaniline loses OH from the molecular ion giving an intense *m/z* 149 ion. Other abundant ions in the mass spectra include *m/z* 119, 105, 104, 77, and 42. For the *p*-isomer, *m/z* 136 is reasonably abundant.

G. Dinitrobenzenes (MW = 168)

Some abundant ions in the mass spectra of the dinitrobenzene isomers include: *m/z* 30, 168, 75, 50, 76, 122, and 92. The *m*- and

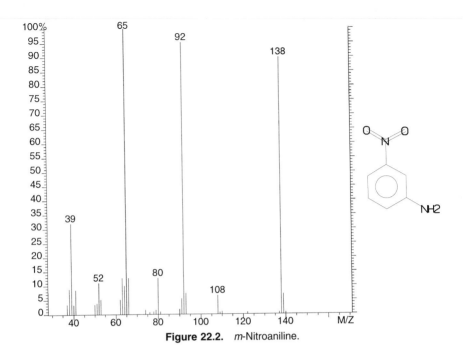

Figure 22.2. *m*-Nitroaniline.

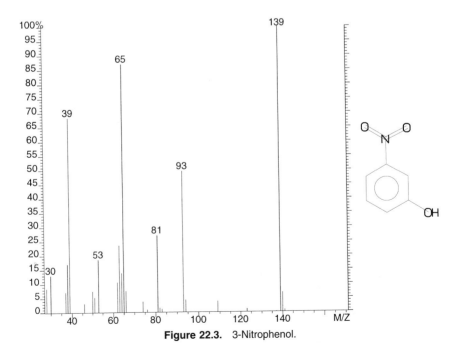

Figure 22.3. 3-Nitrophenol.

p-isomers can be distinguished by the ratio of m/z 76 to m/z 75. In the mass spectrum of the *m*-isomer m/z 76 and m/z 75 have approximately equal intensities, whereas in the *p*-isomer, m/z 75 is more intense than m/z 76. The *o*-isomer can be distinguished from the other isomers by the greater abundance of the m/z 63 and m/z 64 ions relative to the m/z 76 and m/z 75 ions.

H. Chloronitrotoluenes (MW = 171)

Intense ions in the mass spectra occur at m/z 171, 141 (M-NO), 125 (M-NO$_2$), and 113 [(M − (NO + CO)]. If chlorine (m/z 35 and 37) is lost from the molecular ion, then the chlorine atom lies on the alkyl group.

M – Cl   M – Cl
(Not observed)   (Intense ion)

I. Nitronaphthalenes (MW = 173)

The most abundant ions in the mass spectra of the 1- and 2-nitro-naphthalenes are m/z 127, 115, and 173. The 1-isomer can be distinguished from the 2-isomer by the presence of m/z 145 and 143. The m/z 115 ion is more abundant for the 1-isomer.

J. Dinitrotoluenes (MW = 182)

The molecular ions are detectable for all isomers except the 2,3-isomer, with the 3,4- and 3,5-isomers being the most abundant and the 2,6-isomer being the least abundant (2%). Because in the 2,3- and 2,6-isomers nitro groups are ortho to the methyl group, OH is readily lost at the expense of the molecular ion. If the molecular ion (m/z 182) is intense, the unknown mass spectrum is either the 3,4- or the 3,5-isomer. If the m/z 165 peak is very abundant then the mass spectrum represents the 2,3-, 2,4-, 2,5-, or a 2,6-isomer.

K. Dinitroanilines (MW = 183)

The molecular ion is the base peak in the spectra of the dinitroanilines. Other important ions occur at m/z 153, 137, 107, and 91. The m/z 137 ion is very weak in the 2,6-isomer, which is also characterized by the loss of 17 Daltons from the molecular ion.

L. Dinitrophenols (MW = 184)

Dinitrophenols are characterized by abundant molecular ions. The 2,4-isomer has intense fragment ions at m/z 154, 107, 91, and 79. None of these ions are abundant in the 2,6- or 2,5-isomers. The 2,6-isomer has an m/z 126 ion, which is not present in the 2,5-isomer. All of the isomers have m/z 63 and m/z 46 ions.

M. Dinitrochlorobenzenes (MW = 202)

Intense ions occur at m/z 30, 202, 75, 110, 63, 74, 50, and 109. Most of the isomers can be distinguished by the ions at m/z 186, 172, and 156. For example, all of these ions are present in the 2,4-isomer, none in the 3,4-isomer, and 186 (small peak) and 156 in the 2,5-isomer.

N. Dichloronitrotoluenes (MW = 205)

Dichloronitrotoluenes are indicated by the presence of an odd molecular ion with chlorine isotopes showing two chlorine atoms and losses of 30 and 46 Daltons. Again, when the chlorine atoms are on the benzene ring, the loss of chlorine from the molecular ion does not occur. An M − Cl ion indicates that at least one of the chlorines is on the alkyl group.

O. Nitrodiphenylamines (MW = 214)

Abundant ions that are characteristic of the mass spectra of nitrodiphenylamines are m/z 214, 184, and 164.

P. Chlorodinitroanilines (MW = 217)

The mass spectra of chlorodinitroanilines are characterized by m/z 217, 201 (small peak), 187 (M − NO), 171 (M − NO_2), 141 [M − (NO_2 + NO)], and 125 [(M − 2(NO_2)].

Q. Dinitronaphthalenes (MW = 218)

All isomers show abundant molecular ions and m/z 126 and m/z 114 ions, which are characteristic of the naphthalene moiety. In the 2,3- and 1,8-dinitro isomers, the m/z 126 peak is very small. The 2,3-isomer has an abundant m/z 127 ion.

R. Trichloronitrotoluenes (MW = 239)

The presence of three chlorine atoms is easily determined by the isotope ratios. The odd molecular weight shows the presence of nitrogen. The loss of m/z 30 and 46 from the molecular ions shows the presence of the nitro group.

S. Sample Mass Spectra

In Figure 22.2, the abundance of the molecular ion at m/z 138 suggests an aromatic compound. Examining the losses from the molecular ion (M-46 and M-30) shows that it is a nitro compound. Because the molecular ion is of even mass, an even number of nitrogen atoms must be present. Looking up m/z 92 in Part III for possible structures, the following is suspected:

The structure is a nitroaniline. Generally, the *o*-isomer can be detected by the losses of both 16 and 17 Daltons from the molecular ion, whereas the *m*- and *p*-isomers lose only 16 Daltons. This ortho effect applies when amino, hydroxyl, and methyl groups are ortho to the nitro group. In this example, only a 16-Dalton loss is observed. This loss along with m/z 65 (the base peak in the mass spectrum) suggests an *m*- or *p*-isomer. The molecular ion appears at m/z 139 in Figure 22.3. The abundance of this ion suggests an aromatic compound and the odd-mass molecular ion suggests an odd number of nitrogen atoms. Examining the losses from the molecular ion shows the losses of 46, 30, and 16 Daltons. These losses and an m/z 30 ion suggest a nitroaromatic compound. Looking up m/z 93 in Part III suggests a nitrophenol. The presence or absence of an OH group can be determined by preparing a TMS derivative.

Chapter 23

Nitrogen-Containing Heterocyclics

I. GC Separations of Nitrogen-Containing Heterocyclics

A. Capillary Columns

1. Pyridines, 2,3,4-methylpyridines (picolines), and aniline
   30 m Carbowax column + KOH on Carbopack B, 75–150° at 3°/min.

2. Quinolines and acridines
   30 m DB-5 column, 75–275° at 10°/min.

3. Pyrroles, indoles, and carbazoles
   30 m DB-17 column, 75–275° at 10°/min.

4. Benzonitrile, aniline, nitrobenzene, benzo-p-diazine, biphenyl, azobenzene, and dibenzoparadiazine
   30 m DB-225 column, 75–215° at 10°/min.

5. Phenazine and anthracene
   30 m DB-1 column, 100–225° at 6°/min.

6. Imidazoles and benzimidazoles
   30 m DB-5 column, 75–250° at 10°/min.

7. Creatinine-TMS and uric acid-TMS
   30 m DB-1 column, 100–250° at 10°/min.

II. Mass Spectra of Nitrogen-Containing Heterocyclics

A. Mass Spectra of Pyridines, Quinolines, and Acridines

1. Molecular ion: The molecular ion is usually abundant except when long-chain alkyl groups are attached to the ring.

2. Fragment ions: Nitrogen-containing heterocyclics, such as pyridines, quinolines, and acridines, lose HCN and/or H_2CN from their molecular ions.

3. Examples

a. Pyridines

Methylpyridines (picolines) and dimethylpyridines (lutidines) have prominent m/z 65 and m/z 66 ions in their mass spectra. Aniline can be distinguished from picolines by the m/z 78 ion in the mass spectra of picolines. If the alkyl group, R, is attached to the carbon atom adjacent to the nitrogen atom, RCN can be lost easily. Alkylpyridines are characterized by ions at m/z 65, 66, 78, 92, 106, and so forth.

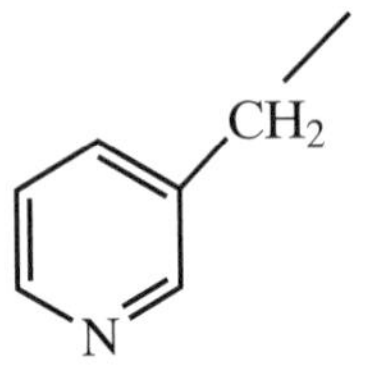

m/z 92 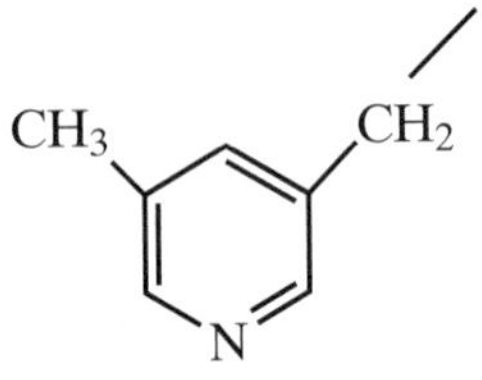   *m/z* 102

b. Quinoline (benzpyridine)

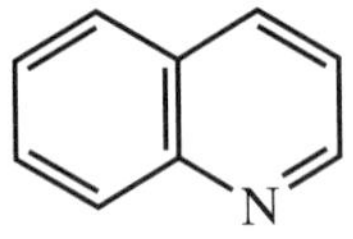

MW = 129

Some characteristic ions in the mass spectra of quinolines include: m/z 102, 128, 156, and so on. Again, as in alkyl pyridines, RCN is lost if the alkyl group is attached to the carbon atom adjacent to the nitrogen atom.

c. Acridine (dibenzpyridine)

MW = 179

The molecular ion is the most abundant ion. Characteristic fragment ions in the mass spectrum occur at *m/z* 89, 90, and 151.

B. Mass Spectra of Pyrroles, Indoles, and Carbazoles

1. Molecular ion: Molecular ions of these heterocyclics are usually abundant except when long chains or tertiary alkyl groups are attached to the ring.

2. Fragment ions: Alkyl groups attached to a carbon atom of the ring (C-alkyl) fragment beta to the ring. A test for the presence or absence of the C-alkyl derivatives can be determined by preparing a TMS derivative. If the TMS derivative cannot be prepared, an alkyl or other group is attached to the nitrogen atom. N-alkyl derivatives fragment beta to the nitrogen atom, accompanied by the rearrangement of a hydrogen atom. For example, N-alkyl derivatives of pyrrole have characteristic fragment ions at *m/z* 81 and 80.

3. Examples

a. Indole (benzopyrrole)

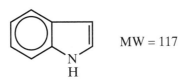

MW = 117

The presence of indole and carbazole can also be determined by their TMS derivatives. Indole-TMS has abundant ions at *m/z* 189, 174, and 73.

b. Carbazole (dibenzopyrrole)

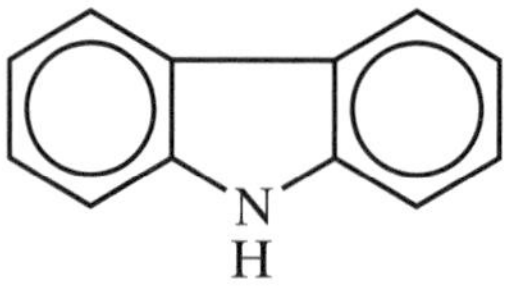

MW = 167

Carbazole-TMS has intense ions at *m/z* 239 (100%), 224 (58%), and 73 (58%). Underivatized alkyl pyrroles have characteristic ions at *m/z* 80, 94, 108, and so forth, while alkyl indoles have characteristic fragment ions at *m/z* 103, 115, 144, and so on.

C. The Mass Spectra of Pyrazines, Quinoxalines, and Phenazines

1. Molecular ions: The molecular ions are observed.

2. Fragment ions: Characteristic fragment ions involve loss of a nitrogen and the adjacent carbon atom (RCN).

3. Examples:

a. Pyrazines 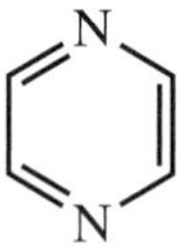 MW = 80

Alkyl pyrazines lose RCN from their molecular ions when the alkyl group is attached to the carbon adjacent to the nitrogen atom.

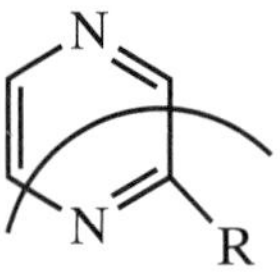

b. Quinoxaline

Quinoxaline gives an intense molecular ion at m/z 130.

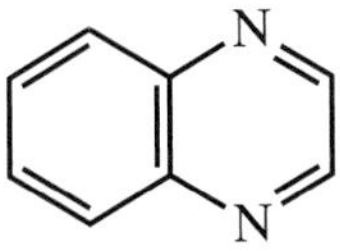

MW = 130

The loss of HCN from the molecular ion yields an intense ion at m/z 103. A second loss of HCN results in an intense ion at m/z 76. In dimethylquinoxaline, loss of acetonitrile occurs:

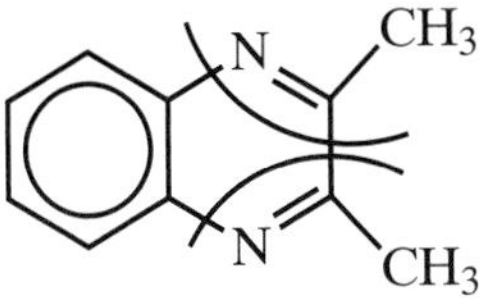

MW = 158

This loss is followed by another loss of acetonitrile, resulting in an intense ion at m/z 76.

c. Phenazine (dibenzopyrazine)

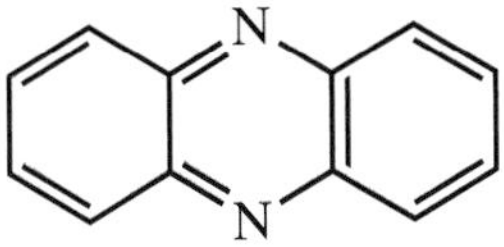

MW = 180

The mass spectrum of phenazine is characterized by ions at m/z 180, 179, 153, 90, and 76.

D. Mass Spectra of Imidazoles and Benzimidazoles

1. Molecular ions: Abundant

2. Fragment ions: All nitrogen-containing heterocyclics lose HCN and/or H_2CN from their molecular ions.

MW = 68 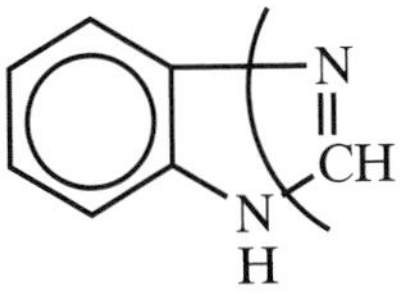   MW = 118

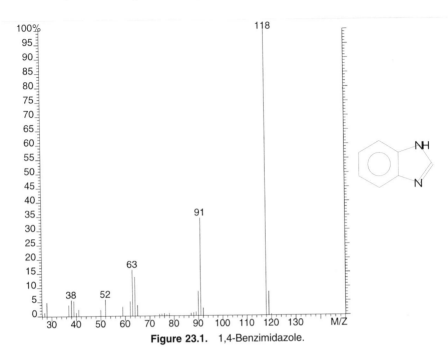

Figure 23.1. 1,4-Benzimidazole.

Imidazole has an abundant molecular ion at *m/z* 68, but loses H and HCN to yield intense ions at *m/z* 67 and *m/z* 41. Benzimidazole loses HCN from the molecular ion (see Figure 23.1). A characteristic fragment ion for alkylbenzimidazoles occurs at *m/z* 132.

E. Creatinine-TMS (MW = 329) and Uric Acid-TMS (MW = 456)

These nitrogen-containing heterocyclics are characterized by the following ions:

Creatinine-TMS   *m/z* 115, 73, 329
Uric acid-TMS   *m/z* 73, 456, 441

Chapter 24

Nucleosides (TMS Derivatives)

I. Derivatization

Add 0.25 ml of DMF (*N,N*-dimethylformamide) and 0.25 ml of TRI-SIL TBT reagent to the sample in a screw-cap septum vial. If TRI-SIL TBT reagent is not readily available, add 0.25 ml of acetonitrile and 0.25 ml of BSTFA reagent instead. Heat at 60° for at least 1 hour for ribonucleosides or for a minimum of 3 hours for deoxyribonucleosides. After cooling to room temperature, inject 1–2 μl of the reaction mixture directly into the GC. The resulting derivatives have been reported to be stable for weeks if capped tightly and refrigerated.[1]

II. GC Separation of Derivatized Nucleosides

A. Capillary column

1. 2′-Deoxyuridine, thymidine, 2′-deoxyadenosine, 2′-deoxycytidine, 2′-deoxyguanosine
   10–30 m DB-17 column, 100–275° at 10°/min; injection port at 280°

III. Mass Spectra of TMS-Nucleosides[2]

Plot *m/z* 103 to determine the elution time of the TMS-nucleosides. Next, determine the molecular weight by identifying the molecular ion

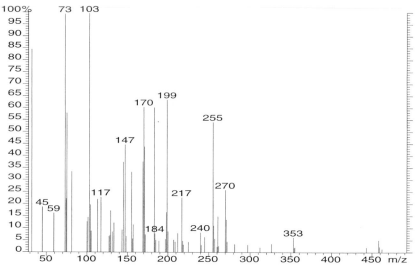

Figure 24.1. TMS derivative of 2′deoxyuridine.

(which is usually observed) associated with an M − 15 ion and often with M − 90, M − 105, and M − 203 ions. The molecular ions of the TMS derivatives of ribonucleosides are 88 Daltons higher than the TMS derivatives of the deoxyribonucleosides. Ions that help identify the base portion of TMS-nucleosides are given in the proceeding text. If the difference between the molecular ion peak and the base peak is 260, the sugar portion is deoxyribose. A difference of 290 represents an *o*-methylribose and of 348 suggests a ribose.

Nucleoside-TMS	MW	*m/z* Values that Indicate the Base	Base
2′-Deoxyuridine	444	169, 184	Uracil
Thymidine	458	183	Thymine (5-methyluracil)
2′-Deoxyadenosine	467	192, 207	Adenine
2′-Deoxycytidine	443	168, 183	Cytosine
2′-Deoxyguanosine	555	280, 295	Guanine

If necessary, the nucleosides can be hydrolyzed to the sugar and the base by heating in formic acid.[2] Kresbach et al.[3] have used pentafluoro-benzylation combined with negative CI for the detection of trace amounts of nucleobases. For fragmentation patterns of TMS 2′-, 3′-, and 5′-deoxynucleosides, see Reimer et al.[4] (See Figure 24.1.)

References

1. Schram, K. H., and McCloskey, J. L. in K. Tsuji, Ed. *GLC and HPLC of Therapeutic Agents.* New York: Marcel Dekker, 1979.
2. Crain, P. F. Chapter 43. in J. A. McCloskey Ed. *Methods in Enzymology* (Vol. 193). San Diego, CA: Academic Press, 1990.
3. Kresbach, G. M., Annan, R. S., Saha, M. G., Giese, R. W., and Vouros, P. *Mass Spectrometric and Chromatographic Properties of Ring-Penta Fluorobenzylated Nucleobases Used in the Trace Detection of Alkyl DNA Adducts.* Proceedings of the 36th Annual Conference of the American Society for Mass Spectrometry, San Francisco, June 5–10, 1988.
4. Reimer, M. L. J., McClure, T. D., and Schram, K. H. *Investigation of the Fragmentation Patterns of the TMS Derivatives of 2′-, 3′-, and 5′-Deoxynucleosides.* Proceedings of the 36th Annual Conference of the American Society for Mass Spectrometry, San Francisco, June 5–10, 1988.

C h a p t e r   2 5

Pesticides

I. Chlorinated Pesticides

A. GC Separations

1. Lindane, heptachlor, aldrin, α- and γ-chlordane, dieldrin, DDT, and similar compounds

   a. 30 m CP-SIL 8 CB* column, 60–300° at 4°/min.

   b. 30 m DB-5 column, 50 (2 min)–140° at 20°/min; 140–300° at 4°/min.*

   c. 15–30 m DB-608 column, 140 (2 min)–240° at 10°/min; 240 (5 min)–265° at 5°/min.

   d. 50 m CPSIL-88 column, 60–225° at 20°/min.

B. Pesticide Extraction Procedures

1. For pesticide extraction procedures pertaining to food samples, refer to U.S. government manuals on pesticide residue analysis.

---

*The DB-5 column may be used, but does not provide enough GC resolution if metabolites are present.

Pesticide Analytical Manual, Volumes I and II, U.S. Depart-
ment of Health and Human Services, Food and Drug Adminis-
tration, U.S. Government Printing Office, Washington, DC,
1994.

2. For pesticide extraction from aqueous samples see: Eisert, R.,
and Leusen, K., *J. Am. Soc. Mass Spectrom.*, *6*, 1119, 1995.

3. The Environmental Protection Agency (EPA) has prepared a
manual of pesticide residue analysis dealing with samples of
blood, urine, human tissue, and excreta, as well as water, air,
soil, and dust.
Manual of Analytical Methods, J. F. Thompson, Ed. Quality
Assurance Section, Chemistry Branch, EPA, Environmental
Toxicology Division, Pesticides, Health Effects Research Laboratory, Research Triangle Park, NC 27711.

4. Pesticide bulletins are available from: Supelco, Inc., Supelco
Park, Bellefonte, PA 16823.

C. Structure of Common Chlorinated Pesticides and Abundant Ions

1. Lindane   **Abundant Ions**

181, 183, 109, 217

$C_6H_6Cl_6$ (MW = 288)

2. Heptachlor

100, 272, 274, 65, 270

$C_{10}H_5Cl_7$ (MW = 370)

3. Aldrin

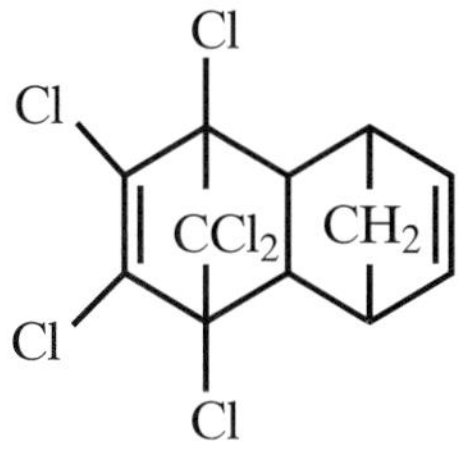

$C_{12}H_8Cl_6$ (MW = 362)

66, 263, 265, 79, 261

4. Chlordane

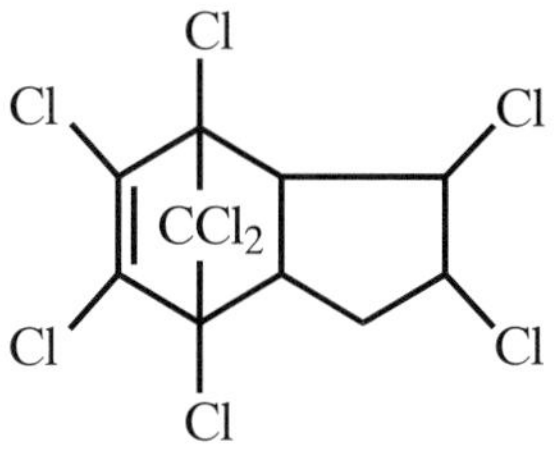

$C_{10}H_6Cl_8$ (MW = 406)

373, 375, 377

The molecular ion for chlordane can be observed in Figure 25.1. Note the pattern for eight chlorine atoms. The most abundant fragment ion is the loss of a chlorine atom at *m/z* 373.

5. Dieldrin

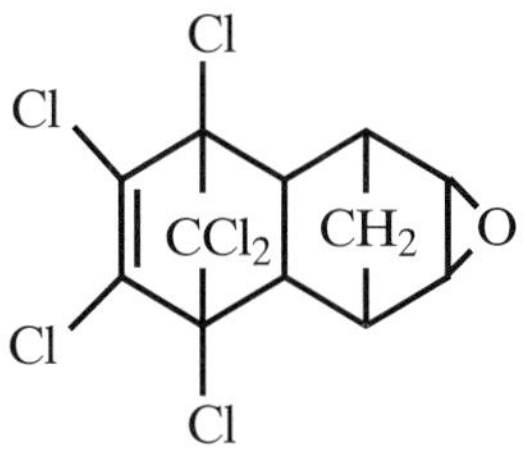

$C_{10}H_8Cl_6$ (MW = 378)

79, 82, 263, 81, 277

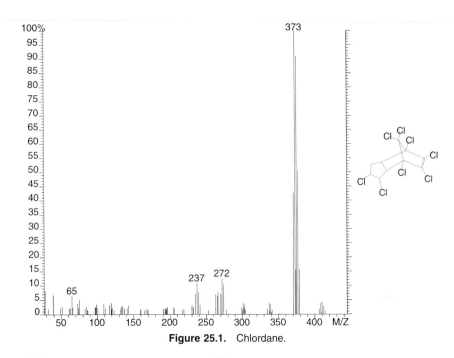

Figure 25.1. Chlordane.

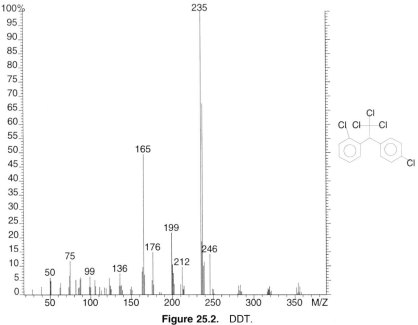

Figure 25.2. DDT.

6. DDT

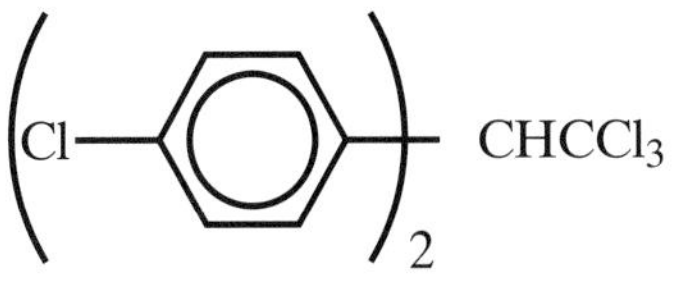

$$C_{14}H_9Cl_5 \text{ (MW = 352)}$$

235, 237, 165, 236, 239

The molecular ion is apparent in the mass spectrum of DDT (Figure 25.2) at *m/z* 352 with the classic isotope pattern for five chlorine atoms (see Appendix 11). The major fragment ion is the loss of CCl_3 at *m/z* 235.

7. Methoxychlor

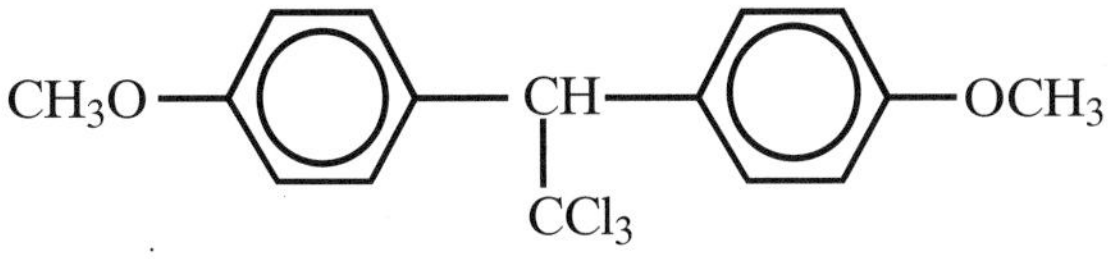

227, 274, 308

$$C_{16}H_{15}Cl_3O_2 \text{ (MW = 344)}$$

II. Organophosphorus Pesticides

A. GC Separations

1. Diazinon, malathion, dimethoate, trichlorofon,* and so on
   30 m DB-5, SPB-5, or Supelco PTE-5 column, 100–300° at 4°/min or 150 (3 min)–250° at 5°/min.

2. Dichlorovos, phorate, dimethoate, diazinon, disulfoton, methyl-parathion, malathion, parathion, azinphos-methyl, azinphos-ethyl, and so on
   50 m CP-SIL 13CB column, 75–250° at 10°/min.

---

*Trichlorofon loses HCl from the molecular ion, producing a spectrum identical to that of dichlorovos. These pesticides can be differentiated by preparing a TMS derivative.

B. Structures of Common Organophosphorus Pesticides and Abundant Ions

1. Dichlorovos*   **Abundant Ions**

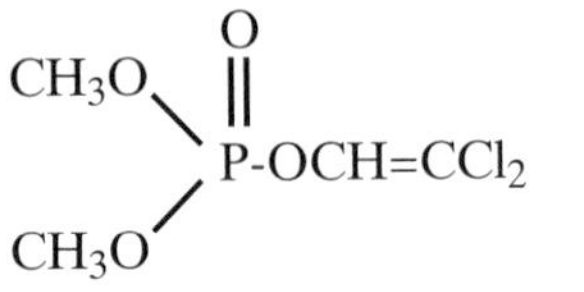

$$CH_3O$$
$$\diagdown$$
$$\qquad P\text{-}OCH\text{=}CCl_2$$
$$\diagup \quad \overset{\displaystyle O}{\overset{\|}{}}$$
$$CH_3O$$

109, 79, 185

$C_4H_7Cl_2O_4P$ (MW = 220)

2. Trichlorofon*

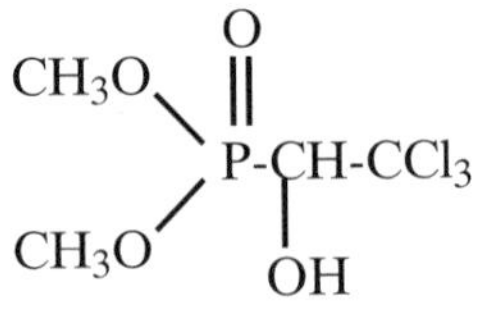

$$CH_3O$$
$$\diagdown \quad \overset{\displaystyle O}{\overset{\|}{}}$$
$$\qquad P\text{-}CH\text{-}CCl_3$$
$$\diagup \qquad |$$
$$CH_3O \qquad OH$$

109, 79, 185

$C_4H_8O_4Cl_3P$ (MW = 256)

3. Phorate (thimet)

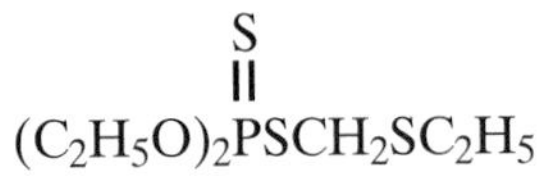

$$\overset{\displaystyle S}{\overset{\|}{}}$$
$(C_2H_5O)_2PSCH_2SC_2H_5$

75, 121, 260, 97

$C_7H_{17}O_2PS_3$ (MW = 260)

---

*Trichlorofon loses HCl from the molecular ion, producing a spectrum identical to that of dichlorovos. These pesticides can be differentiated by preparing a TMS derivative.

4. Dimethoate

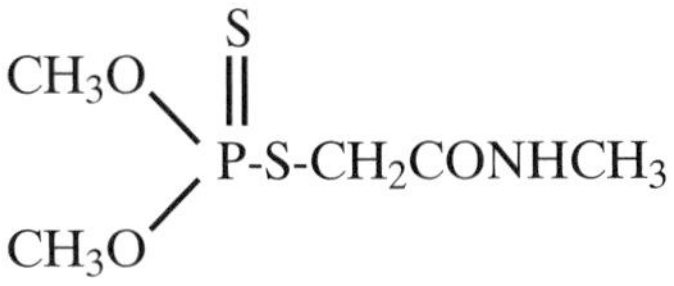

$$CH_3O\diagdown \overset{\overset{\text{S}}{\|}}{P}\text{-S-CH}_2\text{CONHCH}_3$$
$$CH_3O\diagup$$

87, 93, 125

$C_5H_{12}NO_3PS_2$ (MW = 229)

5. Diazinon (dimpylate)

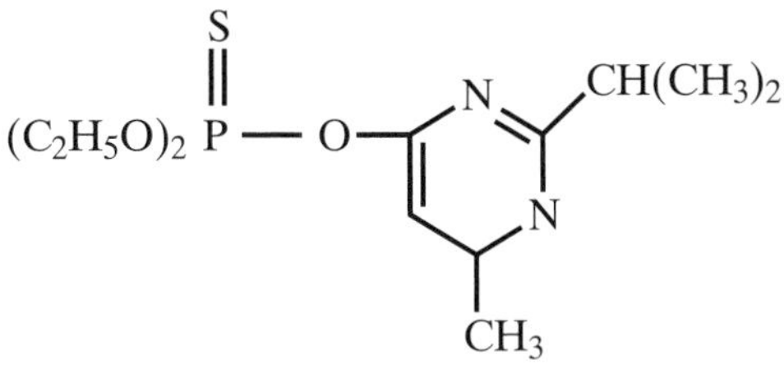

$$(C_2H_5O)_2\ \overset{\overset{\text{S}}{\|}}{P}\text{ — O}$$

with pyrimidine ring: N, CH(CH$_3$)$_2$, N, CH$_3$

179, 137, 152, 199, 304

$C_{12}H_{21}N_2O_3PS$ (MW = 304)

6. Disulfoton

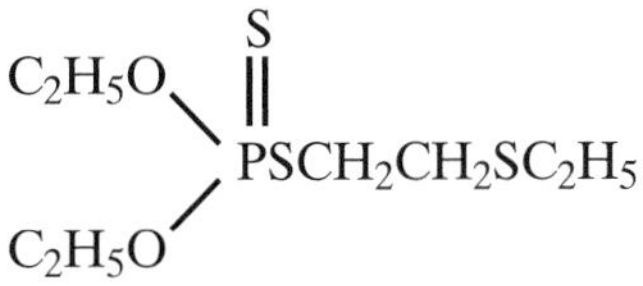

$$C_2H_5O\diagdown \overset{\overset{\text{S}}{\|}}{P}\text{SCH}_2\text{CH}_2\text{SC}_2\text{H}_5$$
$$C_2H_5O\diagup$$

88, 89, 97, 274

$C_8H_{19}O_2PS_3$ (MW = 274)

7. Methyl parathion

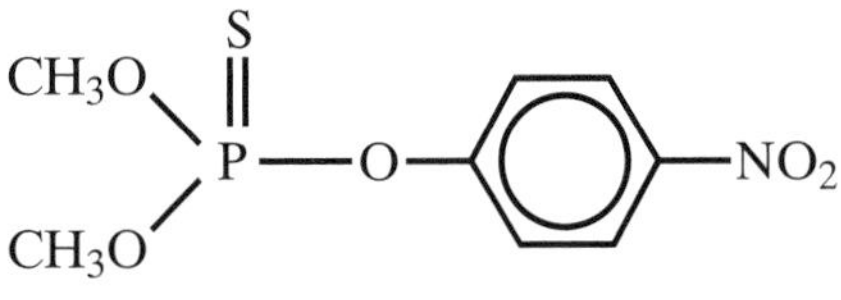

$$CH_3O\diagdown \overset{\overset{\text{S}}{\|}}{P}\text{ — O —}\bigcirc\text{— NO}_2$$
$$CH_3O\diagup$$

109, 263, 125

$C_8H_{10}NO_5PS$ (MW = 263)

8. Malathion

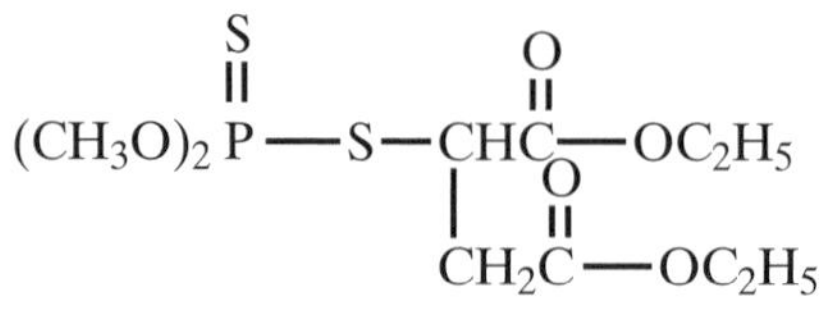

$$(CH_3O)_2 \overset{\overset{\displaystyle S}{\|}}{P}-S-\underset{\underset{\displaystyle CH_2\overset{\overset{\displaystyle O}{\|}}{C}-OC_2H_5}{|}}{CH}\overset{\overset{\displaystyle O}{\|}}{C}-OC_2H_5$$

173, 127, 125, 93, 158

$C_{10}H_{19}O_6PS_2$ (MW = 330)

9. Parathion

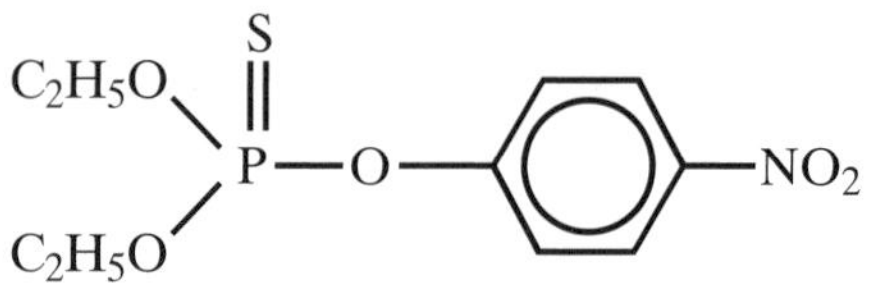

$$\underset{C_2H_5O}{\overset{C_2H_5O}{\diagdown}}\overset{\overset{\displaystyle S}{\|}}{P}-O-\!\!\!\bigcirc\!\!\!-NO_2$$

109, 97, 137, 291

$C_{10}H_{14}NO_5PS$ (MW = 291)

10. Azinphos-methyl

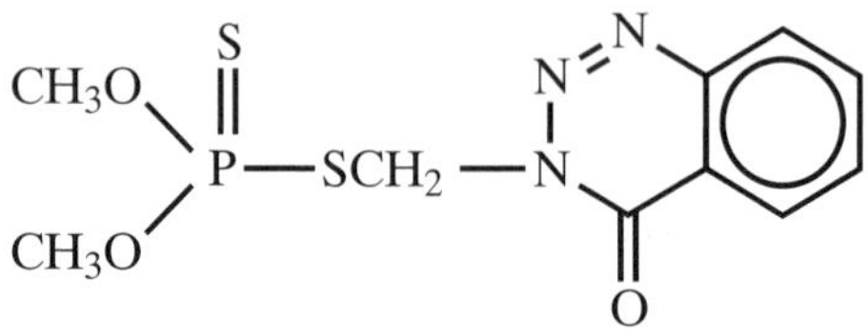

$$\underset{CH_3O}{\overset{CH_3O}{\diagdown}}\overset{\overset{\displaystyle S}{\|}}{P}-SCH_2-N$$

77, 132, 160

$C_{10}H_{12}N_3O_3PS_2$ (MW = 317)

11. Azinphos-ethyl

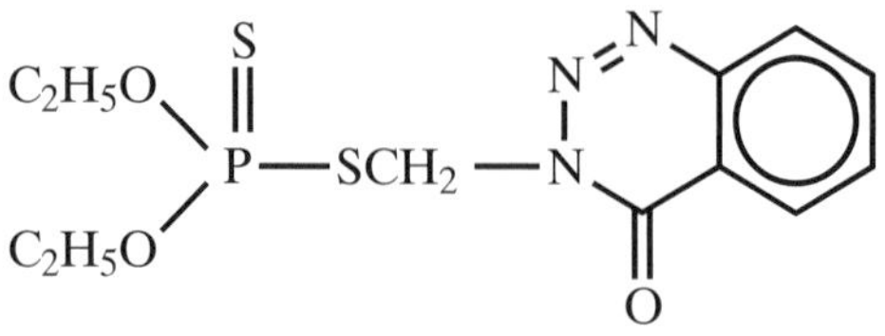

$$\underset{C_2H_5O}{\overset{C_2H_5O}{\diagdown}}\overset{\overset{\displaystyle S}{\|}}{P}-SCH_2-N$$

132, 160, 77, 186

$C_{12}H_{16}N_3O_3S_2P$ (MW = 345)

III. Mass Spectra of Pesticides

If you frequently analyze pesticides, obtain the latest edition of *Mass Spectrometry of Pesticides and Pollutants* (Safe and Hutzinger, Boca Raton, FL, CRC Press). This book, combined with the list of most abundant ions (Table 25.1) and/or a computer library search, will be sufficient to identify most commercial pesticides. Also, see Chapters 17, 26, and 27.

Table 25.1. Ions for identifying pesticides

Base peak	Four next most intense peaks				Compound	Highest m/z peak > 1%
66	263	265	79	261	Aldrin	362
75	121	260	97	—	Phorate (thimet)	260
77	32	160	93	76	Azinphos-methyl	317
79	82	263	81	277	Dieldrin	378
81	100	61	60	59	Methyldemeton	230
87	75	55	—	—	Aldicarb	190
87	93	125	229	—	Dimethoate	229
88	89	97	274	—	Disulfoton	274
97	197	199	314	—	Chlorpyrifus	349
100	272	274	65	270	Heptachlor	370
109	79	185	145	—	Dichlorovos	220
109	79	185	145	—	Trichlorofon	256
109	81	149	99	—	Paraoxon	275
109	97	137	291	—	Parathion	291
109	263	125	—	—	Methyl parathion	263
110	152	81	—	—	Propoxur	209
132	160	77	—	—	Azinphos-ethyl	345
173	127	125	93	158	Malathion	330
179	137	152	199	304	Diazinon (dimpylate)	304
181	183	109	217	—	Lindane	288
227	274	308	—	—	Methoxychlor	344
235	237	165	236	239	DDT	352
373	375	377	—	—	Chlordane	404

C h a p t e r   2 6

Phenols

I. GC Separations of Underivatized Phenols and Dihydroxybenzenes

A. Capillary columns

1. General conditions for separation of phenols
   30 m DB-5 column, 80–150° at 8°/min.

2. Phenol; 2,6-xylenol; *m*-, *p*-, and *o*-cresols; 2,4,6-trimethylphenol; 2,5-xylenol; 2,4-xylenol, 2,3,6-trimethylphenol; 2,3-xylenol; 3,5-xylenol; and 3,4-xylenol
   25 m DB-1701 column, 75 (2 min)–140° at 4°/min.

3. Phenol; 2,4,6-trichlorophenol; *p*-chloro-*m*-cresol; 2-chlorophenol; 2,4-dichlorophenol; 2,4,-dimethylphenol; 2-nitrophenol; 4-nitrophenol; 2,4-dinitrophenol; and pentachlorophenol
   25 m DB-1701 column, 40 (2 min)–210° at 6°/min.

4. Dowtherm impurities: benzene, phenol, naphthalene, and dibenzofuran from diphenyl and diphenyl ether. (Diphenyl and diphenyl ether are not separated under these conditions.)
   30 m DB-225 column, 75–215° at 8°/min.

5. Dihydroxybenzenes, catechol, and resorcinol
   30 m DB-WAX column, 150—230° at 10°/min.

II. Derivatization of Phenols and Dihydroxybenzenes

Add 250 μl of MSTFA or TRI-SIL/BSA formula D to approximately 1 mg of sample in a septum-stoppered vial. (TRI-SIL/BSA formula D contains DMF, which may interfere in the GC separation of some low-boiling TMS derivatives.) Heat at 60° for 15 min.

III. GC Separations of Derivatized Phenols and Dihydroxybenzenes

A. Capillary columns

   1. Catechol-TMS, resorcinol-TMS, and hydroquinone-TMS (MW = 254). (The presence of phosphoric acid interferes with this separation.)
30 m DB-1 column, 60 (5 min)–250° at 4°/min.

   2. TMS derivatives of phenols and dihydroxybenzenes
30 m DB-1 column, 60 (5 min)–250° at 4°/min.

IV. Mass Spectra of Phenols

A. Underivatized Phenols

   1. Molecular ion: The molecular ions are very intense in phenol, methylphenol, and dimethylphenol.

   2. Fragment ions: The losses of 28 and 29 Daltons from the molecular ions are characteristic. Methylphenol can be distinguished from dimethylphenol by comparing the M − 1 and the M − 15 peaks. Methylphenol has an intense M − 1 ion, whereas the M − 15 for dimethylphenol is more abundant. Methylphenol can be distinguished from benzyl alcohol by the m/z 107 ion, which is characteristic of alkylphenols.

B. Sample Mass Spectrum of an Underivatized Phenol

The mass spectrum in Figure 26.1 shows a molecular ion at m/z 122. The abundant fragment ion at m/z 107 can be either of the

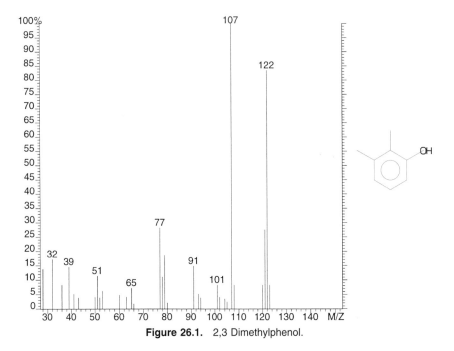

Figure 26.1. 2,3 Dimethylphenol.

following structures as suggested by Part III (R + DB = 4) (see Chapter 2):

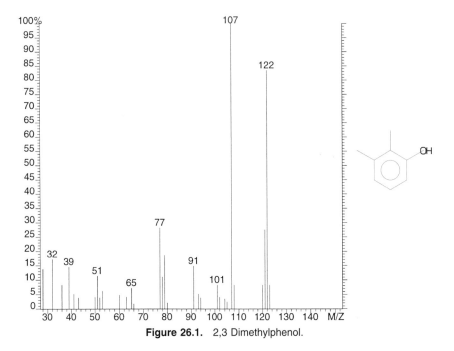

I   II   III

Losses of 28 and 29 Daltons from the molecular ion suggests that the hydroxyl group is attached to the benzene ring, thus eliminating structure I. By preparing a TMS derivative, the presence of the hydroxyl group can be established. However, the position of the aromatic substitution cannot be determined using only mass spectrometry.

C. Phenols as TMS Derivatives

1. Molecular ion: Abundant molecular ions are observed.

2. Fragment ions: Benzyl alcohol-TMS can be distinguished from a cresol-TMS because benzyl alcohol has its base peak at *m/z* 91. The *o*-, *m*-, and *p*-cresol derivatives can be distinguished from each other and from benzyl alcohol by the relative intensities of the *m/z* 91 ion.

TMS-Derivative	Abundant Ions
Phenol-TMS	151, 166
Cresol-TMS	165, 180, 91
Benzyl alcohol-TMS	91, 165, 135, 180
Xylenol-TMS (6 isomers)	179, 194, 105

If a phenol is suspected after running the sample without derivatization, perform a run using the TMS derivative to determine the presence and number of hydroxyl groups.

V. Aminophenols (see Chapter 8)

VI. Antioxidants*

A. GC Separation of Antioxidants

2,6-Di-*tert*-butyl-4-methylphenol (BHT), 2,4-di-*tert*-butylphenol, and 6-*tert*-butyl-2,4-dimethylphenol
30 m Supelcowax-10 column, 50–200° at 5°/min.

B. Mass Spectra of Antioxidants

Antioxidant	Abundant Ions
2,6-Di-*tert*-butyl-*p*-cresol (BHT) (IONOL)	205, 57, 220
2,4-Di-*tert*-butylphenol	191, 57, 192, 206
6-*Tert*-butyl-2,4-dimethylphenol	163, 135, 178

---

*For further information on polymer additives including antioxidants, see: Cortes, J. H., Bell, B. M., Pfeiffer, C. D., and Graham, J. D. *J. Microcolumn Separations*, *1*, 278, 1989; and Asamoto, B., Young, J. R., and Citerin, R. J., *Anal. Chem.*, *62*, 61, 1990.

C h a p t e r   2 7

Phosphorus Compounds

I. GC Separations

A. Capillary Columns

1. Hexamethylphosphoramide, pentamethylphosphoramide, tetra-methylphosphoramide, trimethylphosphoramide, and the metabolite that was postulated to be

$$(CH_3)_2N \diagdown \underset{\underset{(CH_3)_2N \diagup}{}}{\overset{\overset{O}{\|}}{P}} - N - \overset{CH_3}{\underset{|}{C}} = O \diagup^H$$

30 m FFAP-DB column, 80–220° at 12°/min or 30 m CW20M column, 60–200° at 10°/min.

2.

$CH_2CH_2CH_2PO_3(C_2H_5)_2$ and $PhOCH_2CH_2P(CH_3)_2$

30 m DB-1 column, 70–220° at 10°/min.

205

3. Phosphorus Pesticides (See Chapter 25)
   Diazinon, malathion, and similar compounds

   a. 30 m DB-5 column, 150 (3 min)–250° at 5°/min or 100–300° at 4°/min.

   b. 50 m CP-SIL 13CB column, 75–250° at 10°/min.

   c. 30 m CP-SIL 7 column, 190–240° at 4°/min.

II. Mass Spectra of Phosphorus Compounds

A. Alkyl Phosphites and Alkyl Phosphonates

1. General formulas: $H\overset{\overset{\displaystyle O}{\|}}{P}(OR)_2$ and $R_1\overset{\overset{\displaystyle O}{\|}}{P}(OR_2)_2$

2. Characteristic fragment ions:

$$R_1\overset{\overset{\displaystyle O}{\|}}{P}(OR_2)_2 \longrightarrow \left[R_1-P\overset{\diagup OH}{\underset{\diagdown OH}{-OH}}\right]^+$$

If R_1 = H, m/z 83 is observed.
If R_1 = CH_3, m/z 97 is observed.
If R_1 = C_2H_5, m/z 111 is observed.
If R_1 contains a γ-hydrogen, then the McLafferty rearrangement occurs.

B. Alkyl Phosphates

1. General formula: $(RO)_3PO$

2. Characteristic fragment ions:

$(C_2H_5O)_3PO$ → $(C_2H_5O)_2P(OH)_2$ → $C_2H_5OP(OH)_3$ → $P(OH)_4$
m/z 182 m/z 155 m/z 127 m/z 99

C. Phosphoramides (Hexamethylphosphosphoramide)

1. Formula: $[(CH_3)_2N]_3PO$

2. Molecular ion: The molecular ion is easily observed.

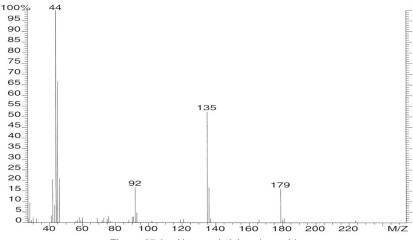

Figure 27.1. Hexamethylphosphoramide.

3. Characteristic fragment ion: A very intense fragment ion is observed at m/z 135 corresponding to $[(CH_3)_2N]_2PO$ (see Figure 27.1).

D. Mass Spectra of Phosphorus Pesticides (See Chapter 25)

Chapter 28

Plasticizers and Other
Polymer Additives
(Including Phthalates)

I. GC Separations

A. Capillary Columns

1. Triethylcitrate (MW = 276), dibutylphthalate (MW = 278), dibutylsebacate (MW = 314), acetyltributylcitrate (MW = 402), trioctylphosphate (MW = 444), di-(2-ethylhexyl)adipate (MW = 370), di-(2-ethylhexyl)phthalate (MW = 390), and di-(*n*-decyl)phthalate (MW = 444)
   15 m DB-1 column, 75–275° at 10°/min.

2. Dimethylphthalate (MW = 194), diethylphthalate (MW = 222), butylbenzylphthalate (MW = 312), di-(2-ethylhexyl)phthalate (MW = 390), and dioctylphthalate (MW = 390)
   15 m DB-1 column, 150–275° at 15°/min.

3. BHT (MW = 220), Tinuvin P (MW = 225), acetyltributylcitrate (MW = 402), butylbenzylphthalate (MW = 312), stearamide (MW = 283), eurcylamide (MW = 337), and Irganox 1076 (MW = 530)
   15 m DB-1 column, 150–300° at 15°/min.

4. Dimethyl terephthalate impurities
   acetone (MW = 58), benzene (MW = 78), toluene (MW = 92), xylene (MW = 106), methylbenzoate (MW = 150), CHO

$(C_6H_4)CO_2CH_3$ (MW = 164), $NC-(C_6H_4)CO_2CH_3$ (MW = 161), methyltoluate (MW = 150), *p*-nitromethylbenzoate (MW = 181), dimethylterephthalate (MW = 194), methyl DMT (MW = 208), $CH_3OC(O)-C_6H_4-CH(OCH_3)_2$ and isomer (MW = 210), and $CH_3OC(O)-C_6H_4-C_6H_3-(CO_2CH_3)_2$ (MW = 328)
30 m DB-1 column, 100–250° at 10°/min.

II. Mass Spectra

Phthalate   Isophthalate   Terephthalate

A. Phthalates, Isophthalates, and Terephthalates

If the MW is not m/z 166 and there is an intense ion at m/z 149, this suggests an ester of a benzene dicarboxylic acid. If m/z 149 is the most intense ion, the mass spectrum represents a phthalate, where R is an ethyl group or larger (see Figure 28.1). The most intense ions in the mass spectra of isophthalates and terephthalates are $[M - OR]^+$.

$m/z = 149$

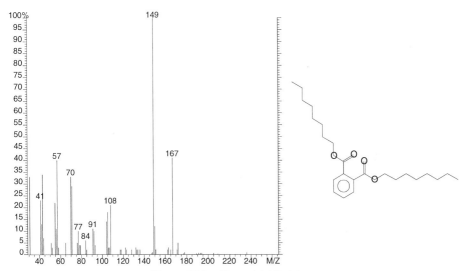

Figure 28.1. Di-*n*-octylphthalate.

Important ions to examine (if present) are the following:

- $M^{+\cdot}$ (small or nonexistent when R is C_5 or higher)
- $[M-(R-H)]^{+\cdot}$ or $[M-(R-2H)]^{+}$
- $[M-OR]^{+}$
- R^{+} (may be only 5%)

1. Dimethyl phthalate (*m/z* 194)

   119 (0.5%), 120 (1.9%), 136 (6%) = M-(COOCH₃)
   163 (100%) = M-(OCH₃)
   194 (11%) = M

2. Dimethyl isophthalate (*m/z* 194)

   119 (4%), 120 (6%), 136 (24%), 163 (100%), 194 (24%)

3. Dimethyl terephthalate (*m/z* 194)

   119 (14%), 120 (23%), 135 (20%), 163 (100%), 194 (25%)

4. Diethyl phthalate (*m/z* 272)

   149 (100%) = M-(OC₂H₅+C₂H₄)
   177 (28%) = M-(OC₂H₅)
   194 (20%) = M-(C₂H₄)
   222 (3.4%) = M

5. Diethyl terephthalate (*m/z* 222)

   149 (45%), 166 (28%) = M-2(C_2H_4)
   177 (100%)
   194 (21%)
   222 (11%)

6. Di-*n*-propylphthalate (*m/z* 250)

   43 (6.6%), 149 (100%), 209 (8%), 250 (0.33%) = M

7. Di-*n*-butylphthalate (*m/z* 278)

   57 (5%), 149 (100%), 223 (5.6%), 278 (0.78%) = M

8. Di-*n*-octylphthalate (*m/z* 390)

   57 (36%), 71 (23%), 149 (100%), 167 (34%), 390 (1%) = M

B. Other Additives

Excellent discussions of the mass spectra of polymer additives are given by Cortes, H. J., Bell, B. M., Pfeiffer, C. D., and Graham, J. D., *J. Microcolumn Separations*, *1(6)*, 278–288, 1989; and Asamoto, B., Young, J. R., and Citerin, R. J., *Anal. Chem.*, *62*, 61, 1990.

Chapter 29

Prostaglandins
(MO-TMS Derivatives)

I. Derivatization (MO-TMS)

Evaporate the sample extract to dryness with clean, dry nitrogen. Store the dried extract at dry-ice temperatures until the sample is derivatized (otherwise the prostaglandin Es [PGEs] and the hydroxy-PGEs may convert to prostaglandin As [PGAs] prostaglandin Bs [PGBs], and so forth). Add 0.25 ml of MOX reagent to the dried extract and let the reaction mixture stand at room temperature overnight. Evaporate to dryness with clean, dry nitrogen. Add 0.25 ml of MSTFA or BSTFA reagent. Let the reaction mixture stand at room temperature for at least 2 hours. Evaporate the excess reagent with flowing nitrogen gas. Dissolve the residue in a minimum amount of hexane for GC/MS analysis.

II. GC Separation of Derivatized Prostaglandins

A. PGA_1-MO-TMS, PGB_1-MO-TMS, PGB_2-MO-TMS, PGE_1-MO-TMS, PGE_2-MO-TMS, PGF_1a-TMS, 6-ketoPGF_1-MO-TMS, PGF_2a-TMS, and thromboxane B_2-TMS

1. 30 m DB-17 column, 175–270° at 8°/min; injection port at 280°.

2. 30 m DB-1701 column, 100 (2 min)–275° at 5°/min; injection port at 280°.

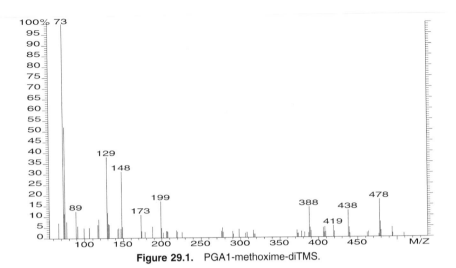

Figure 29.1. PGA1-methoxime-diTMS.

3. 30 m DB-225 column, 100–225° at 15°/min; injection port at 280°.*

III. Mass Spectra of MO-TMS Derivatives of Prostaglandins

A. Molecular Ion: The molecular ions of the MO-TMS derivatives are usually of low intensity but are detectable (see Figures 29.1 and 29.2).

B. Characteristic Fragment Ions: Characteristic fragment ions are M − 15 and M − 31, with the M − 31 ion being more abundant. The M − 31 ion is intense only for ketoprostaglandins. By plotting the masses of the proceeding fragment ions, the presence or absence of a particular prostaglandin can be determined even though complete GC resolution may not be obtained.

*These GC conditions are suitable for analyzing many prostaglandins, thromboxanes, leuko-trienes, and other metabolites of arachidonic acid, such as the hydroxyeicosatetraenoic (HETE) acids. However, the 5-, 12-, and 15-HETE isomers are difficult to separate using GC methods. Sometimes the methyl ester-TMS derivatives provide a better GC separation, or for ketoprostaglandins, the MO-methyl ester-TMS derivatives often give a better separation.

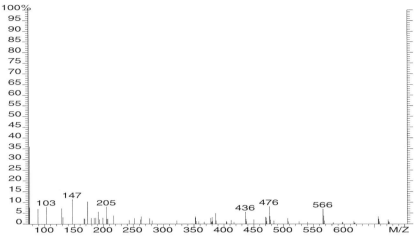

Figure 29.2. 6-Keto-PGF1-methoxime-tetraTMS.

Prostaglandin	Characteristic Abundant Ions
PGA_1-MO-TMS (MW = 509)	*m/z* 388, 419, 438, 478
PGB_1-MO-TMS (MW = 509)*	*m/z* 388, 438, 478, 494
PGB_2-MO-TMS (MW = 507)	*m/z* 386, 476, 492, 507
PGE_1-MO-TMS (MW = 599)	*m/z* 426, 478, 528, 568, 584

Prostaglandin	Most Abundant Ions
PGE_2-MO-TMS (MW = 597)	*m/z* 436, 566, 582, 476
PGF_1a-TMS (MW = 644)†	*m/z* 367, 368, 438, 483
6-KetoPGF$_1$-MO-TMS (MW = 687)	*m/z* 436, 476, 566, 656
PGF_2a-TMS (MW = 642)	*m/z* 391, 462, 481, 552
Thromboxane B_2-TMS (MW = 658)	*m/z* 211, 301, 387
12-HETE-TMS (MW = 464)	*m/z* 213, 324, 449

*PGB_1 can be distinguished from PGA_1 by the more intense *m/z* 478 ion relative to the other fragment ions in the mass spectrum of PGB_1-MO-TMS.

†In the mass spectra of TMS derivatives of nonketoprostaglandins, the molecular ion is usually not observed. Generally, the fragment ions are M − 15, M − 71, M − 90, M − 161, and M − 180.

Chapter 30

Solvents and Their Impurities

There are generally two types of analyses that are requested with reference to solvents. The first is the identification of residual solvents in products, and the second is the identification of impurities in common industrial solvents. Certain GC conditions have been found to separate most of the common solvents. Always examine the mass spectra at the front and back of the GC peaks to determine if they are homogeneous. Also remember that isomers may not be detected by this approach if they are not separated.

I. GC Separations of Industrial Solvent Mixtures

A. Capillary Columns

1. Acetone, THF, methanol, ethyl acetate, isopropyl acetate, isopropyl alcohol, ethanol, *n*-propyl alcohol, toluene, and 2-methoxyethanol

   a. 30 m DB-FFAP column or 30 m HP-FFAP column, 60–200° at 6°/min.

   b. 25 m DB-WAX column or 25 m CP-WAX 52CB column, 50 (2 min)–65° at 1°/min; 65–150° at 10°/min.

2. Ethanol, acetonitrile, acetone, diethyl ether, pentane, ethyl acetate, and hexane
   25 m Poraplot Q column, 60–200° at 6°/min.

3. Benzene, propyl acetate, allyl acetate, 1-pentanol, cyclohexanone, cyclohexanol, dicyclohexyl ether, cyclohexyl valerate, butyric acid, valeric acid, caproic acid, 1,5-pentanediol, dicyclohexyl
succinate, and dicyclohexyl glutarate
30 m DB-FFAP column, 60–200° at 6°/min.

B. Packed Columns

1. Methanol, ethanol, acetone, methylene chloride, methyl acetate,
THF, methyl cellosolve, ethyl cellosolve, butyl cellosolve,
n-xylene, *p*-xylene, and *o*-xylene
10-ft 3% SP-1500 column on 80-120 MESH Carbopack*
60–225° at 4°/min. Run for 50 min.

2. Acetaldehyde, methanol, ethanol, ethyl acetate, *n*-propyl alcohol, isobutyl alcohol, acetic acid, amyl alcohol, and isoamyl alcohol
10-ft Carbowax 20M column on Carbopack B, 70–170° at 5°/min.

II. GC Separations of Impurities in Industrial Solvents

A. Capillary Columns

1. Diaminotoluene impurities: dichlorobenzene, aminotoluene, nitrotoluene isomers, and dinitrotoluene isomers
30 m DB-17 column, 100–250° at 8°/min.

2. 3,4-Dichloroaniline impurities: aniline, chloroaniline, dichloro-
aniline isomers, trichloroaniline isomers, tetrachloroaniline, tet-
rachloroazobenzene, and pentachloroazobenzene
30 m DB-1 column, 100–280° at 10°/min.

3. Diethylene glycol impurities: acetaldehyde, 2-methyl-1,3-dioxo-
lane, dioxane, and diethylene glycol
30 m Nukol column, 50–220° at 8°/min.

4. *N,N*-dimethylacetamide (DMAC) impurities: *N,N*-dimethylace-
tonitrile, *N,N*-dimethylformamide, DMAC, *N*-methylacetamide,
and acetamide
60 m DB-WAX column, 60–200° at 7°/min or 3 m Tenax-GC
packed column, 100–120° at 3°/min.

5. Ethanol impurities: ethyl acetate, methanol, *n*-propyl alcohol,
isobutyl alcohol

*This is the only column we have found that separates the cellosolves (monoalkyl ethers of
ethylene glycol) and the other solvents including the xylenes. (See Supelco GC Bulletin 824.)

25 m CP-WAX 57 column for glycols and alcohols (Chrompack catalog no. 7615), 50 (5 min)–180° at 8°/min.

6. Isopropyl alcohol impurities: ethanol, *n*-propyl alcohol, and *t*-butyl alcohol
   30 m NUKOL column, 45–200° at 6°/min.

7. *t*-Butyl alcohol, *n*-butyraldehyde, methyl ethyl ketone, 2-methylfuran, tetrahydrofuran, 4-methyl-1,3-dioxolane, 2-methyl THF, 3-methyl THF, and tetrahydropyran
   30 m DB-1 column at 60°

8. Toluene impurities: benzene, ethylbenzene, and *o*-, *m*-, and *p*-xylenes
   25 m DB-WAX column, 50 (2 min)–180° at 5°/min.

B. Packed Columns

Tetrahydrofuran (THF) impurities: acetone, acrolein, 2,3-DHF, butyraldehyde, isopropyl alcohol, THF, 1,3-dioxolane, 2-methyl THF, benzene, and 3-methyl THF
10-ft 3% SP-1500 column on Carbopack B, 60–200° at 6°/min.

III. Mass Spectra of Solvents and Their Impurities

Solvents and their impurities represent a wide class of compound types; therefore, a discussion of common mass spectral features is meaningless. However, most of the mass spectra are listed in computer library search programs and "The Eight Peak Index."

Chapter 31

Steroids

I. GC Separation of Underivatized Steroids

A. Animal or Plant Steroids

30 m DB-17 column, 225–275° at 5°/min; injection port at 290–300°. (Dissolve the dried extract in the minimum amount of methylene chloride or toluene.)

II. Derivatization of Steroids

A. Preparation of TMS Derivatives: Add 0.25 ml of TRI-SIL TBT reagent (the only reagent we found to react with the hydroxyl group in position 17) and 0.25 ml of pyridine to the dried extract. Cap tightly and heat at 60° for 1–12 hours or overnight.

B. Preparation of Methoxime (MO)-TMS Derivatives (especially for hydroxyketosteroids, which may decompose under the given GC conditions unless the MO-TMS derivatives are prepared.): Add 0.25 ml of methoxime hydrochloride in pyridine to the dried extract. Let stand for 3 hours at 60° or overnight at room temperature. Evaporate to dryness with clean, dry nitrogen. Add 0.25 ml of TRI-SIL TBT and 0.25 ml of pyridine to the dried reaction mixture. Cap tightly and heat at 60° for a few hours or overnight at room temperature.

III. GC Separation of Derivatized Steroids

A. TMS Derivatives

1. 30 m DB-17 column, 225–275° at 5°/min.

2. 15 m CP-SIL 5 column or 15 m CP-SIL 8 column, 200–300° at 15°/min.

B. MO-TMS Derivatives

1. 30 m DB-17 column, 100–200° at 15°/min, then 200–275° at 5°/min; injection port temperature at 280–300°.

2. 30 m CP-SIL 5 column 100–210° at 20°/min, then 210–280° at 2°/min.

IV. Mass Spectra of Underivatized Steroids

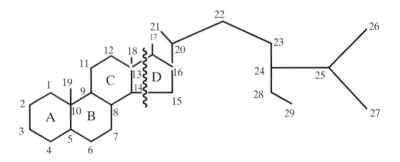

A. Molecular Ion: Because of low volatility, most steriods are derivatized before analysis by GC/MS. Molecular ions are usually observed for steriods sufficiently volatile to be analyzed underivatized by GC/MS (see Figure 31.1) Some important steroids in urine include estrone, estradiol, estriol, pregnanediol, and 17-ketosteroids, which can be analyzed by GC/MS as the TMS or the MO-TMS derivatives. The plant steroids, such as camposterol, ergosterol, stigmasterol, cholestanol, and sitosterol, are generally analyzed as the TMS derivatives.

B. Characteristic Fragment Ions: The most characteristic cleavage is the loss of carbons 15, 16, and 17, together with the side chain and one additional hydrogen. It is possible to determine the elemental composition of the side chains of steroids by the difference in the mass between the molecular ion and an intense ion more than 15

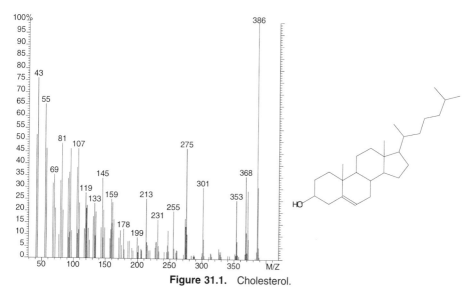

Figure 31.1. Cholesterol.

Daltons below the molecular ion. Typically, this ion corresponds to M − (R + 42), where 42 is C_3H_6. If there is no side chain, as in the cases of estrone, estradiol, and estriol, the substitution on the D-ring can be determined using the following losses from the molecular ions:

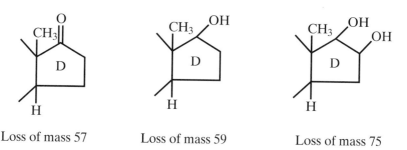

Loss of mass 57   Loss of mass 59   Loss of mass 75

V. Mass Spectra of TMS Derivatives of Steroids

Plot mass 73 to determine if a TMS derivative was prepared and which GC peak(s) to examine.

A. Molecular Ion: Generally, the molecular ion is observed, but not always, as in the case of pregnanediol-TMS (MW = 464). Again, if two high-mass peaks are not observed that are 15 Daltons apart,

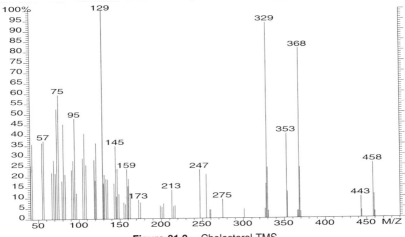

Figure 31.2.   Cholesterol-TMS.

then add 15 Daltons to the highest mass observed to deduce the molecular ion.

B. Fragment Ions: Common losses from the molecular ions include 90 and 105 Daltons. (See Figure 31.2 for the mass spectrum of the TMS derivative of cholesterol.)

VI. Mass Spectra of MO-TMS Derivatives

A. Molecular Ion: The molecular ions of MO-TMS derivatives are generally more intense than those of the TMS-only derivatives.

B. Fragment Ions: The M − 31 fragment ion is characteristic and is more intense than the M-15 ion. If two high-mass ions are observed that are 16 Daltons apart, add 31 Daltons to the more intense ion and 15 Daltons to the highest mass ion to deduce the molecular ion. By subtracting 29 Daltons for each keto group and 72 Daltons for each hydroxyl group, the original molecular weight of the steroid can be determined. For example, in the case of androsterone-MO-TMS, the molecular ion occurs at m/z 391. By subtracting 101 (29 + 72) Daltons, the original molecular weight was m/z 290.

Chapter 32

Sugars (Monosaccharides)

I. GC Separation of Derivatized Sugars

A. Monosaccharides

    1. Preparation of the TMS derivative: Add 0.5 ml of TRI-SIL Z reagent (trimethylsilylimidazole in pyridine) to 1–5 mg of the sample. (This derivatizing preparation does not react with amino groups and tolerates the presence of water.) Heat in a sealed vial at 60° until the sample is dissolved. An alternate method is to let the reaction mixture stand at room temperature for at least 30 minutes (or overnight). This procedure is not appropriate for amino sugars.

    2. GC separation of TMS derivatives: arabinose, fucose, xylose, mannose, galactose, α-glucose, and β-glucose
30 m DB-1 column, 60–250° at 8°/min.

B. Amino Sugars (Glucosamine and Galactosamine as TMS Derivatives)

    1. Preparation of TMS derivative: To derivatize the amino sugars as well as the nonamino sugars, substitute TRI-SIL TBT or TRI-SIL/BSA (Formula P) reagent for TRI-SIL Z and follow the procedure given in Section I,A,1.

2. GC separations of amino sugars as TMS derivatives
   30 m DB-1701 column, 70–250° at 4°/min. (Depending on the amounts present, complete GC separation may not be achieved.)

C. Sugar Alcohols (as Acetates)

   1. Preparation of the acetate derivative: Evaporate the extract to dryness. Add 50 μl of three parts acetic anhydride and two parts pyridine. Heat at 60° for 1 hour. Evaporate to dryness with clean, dry nitrogen and dissolve the residue in 25 μl of ethyl acetate.

   2. GC separation of the acetate derivatives: rhamnitol, fucitol, ribitol, arabinitol, mannitol, galacitol, glucitol, and inositol
      30 m CP-Sil 88 column at 225° for 60 min.

D. Reduced Sugars (as Acetates)

   1. Preparation of the acetate derivative: Concentrate the aqueous mixture of saccharides to approximately 0.5 ml in a 20–50 ml container. Reduce the saccharides by adding 20 mg of sodium borohydride that has been dissolved carefully into 0.5 ml of water and let the reducing mixture stand at room temperature for at least 1 hour. Destroy the excess sodium borohydride by adding acetic acid until the gas evolution stops. Evaporate the solution to dryness with clean nitrogen. Add 10 ml of methanol and evaporate the solution to dryness. Acetylate with 0.5 ml (three parts acetic anhydride and two parts pyridine) overnight. Evaporate to a syrupy residue and add 1 ml of water. Evaporate again to dryness to remove the excess acetic anhydride. Dissolve the residue in 250 μl methylene chloride.

   2. GC separation of reduced and acetylated sugars
      30 m CP-Sil 88 column at 225°.

II. Mass Spectral Interpretation

A. Mass Spectra of TMS Derivatives of Sugars

   1. Molecular weight: In general, to deduce the molecular weight of the TMS derivative of sugars, add to 105 the highest mass observed.

      • 333.1374 is the highest mass observed for arabinose, ribose, ribulose, xylose, lyxose, and xylulose.
      • 347.1530 is the highest mass observed for fucose and rhamnose.
      • 435.1875 is the highest mass observed for sorbose, allose, altrose, galactose, gulose, idose, and mannose.

2. Fragmentation: DeJongh et al.[1] have described the fragmentation of the TMS derivatives of sugars. Comparison of GC retention times together with the mass spectra is sufficient to identify the sugars. The mass spectra suggest certain structural features. For instance, m/z 191, 204, and 217 suggest a TMS hexose. If m/z 204 is more abundant than m/z 217, the hexose is the pyranose form, but if m/z 217 is most abundant then it is a furanose (Figure 32.1). Aldohexoses can be differentiated from ketohexoses by the ion at m/z 435 for aldohexoses versus m/z 437 for ketohexoses.

m/z	**Ion Structure**
191	$(CH_3)_3$ Si O CH $=$ O Si $(CH_3)_3$ ⌉$^+$
204	$(CH_3)_3$ Si O CH $=$ CH O Si $(CH_3)_3$ ⌉$^{+\cdot}$
217	$(CH_3)_3$ Si O CH $=$ CH - CH $=$ O Si $(CH_3)_3$ ⌉$^+$

B. Sample Mass Spectrum

C. Mass Spectra of Amino Sugars as TMS Derivatives

1. Molecular ion: By plotting out the molecular ion mass and mass 362.1639, the presence of these amino sugars can be determined.

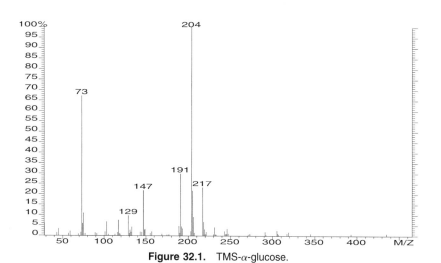

Figure 32.1. TMS-α-glucose.

	Highest Mass Observed
Glucosamine	393.1643
Galactosamine	467.2375

2. Fragmentation: The TMS derivatives of amino sugars have their base peak at mass 131.0765.

D. Mass Spectra of Sugar Alcohols as Acetates

1. Molecular ion: Chemical ionization using ammonia as reagent gas establishes the molecular weights of sugar acetates.

2. Fragmentation: The acetates have intense ions at *m/z*s 43, 103, and 145. They also appear to lose 42, 59, 60, and 102 Daltons from their molecular ions.

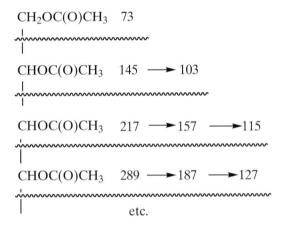

The mass spectrum of alditol acetates are easy to interpret as they fragment at each C—C bond as shown.

Reference

1. DeJongh, D. C., Radfon, T., Hribar, J. D., Hanessian, S., Biebar, M., Dawson, G., Sweeley, C. C. Analysis of derivatives of carbohydrates by GC/MS. *J. Am. Chem. Soc.*, *91*, 1728, 1969.

Chapter 33

Sulfur Compounds

I. GC Separations

A. Capillary Columns

1. Hydrogen sulfide, carbonyl sulfide, sulfur dioxide, and methyl mercaptan
   25 m GS-Q column or 25 m Poraplot Q column, 50–120° at 5°/min.

2. Hydrogen sulfide, methane, carbon dioxide, ethane, and propane
   25 m Poraplot Q column or 25 m GS-Q column at 60°.

3. Carbon dioxide, carbonyl sulfide, hydrogen cyanide, propylene, butadiene, and furan
   25 m GS-Q column or 25 m Poraplot Q column, 60–200° at 6°/min.

4. Carbonyl sulfide, carbon disulfide, tetrahydrofuran, and toluene
   25 m DB-1 column, room temperature to 200° at 10°/min.

5. Methane, ethene, ethane, propene, acetaldehyde, methyl formate, butene, acetone, furan, dimethyl sulfide, isoprene, isobutyraldehyde, diacetyl, methylfuran, and isovaleraldehyde
   25 m Poraplot Q column or 25 m GS-Q column, 60–200° at 8°/min.

6. 2-Propanethiol, 1-propanethiol, 1-methyl-1-propanethiol, 2-methyl-1-propanethiol, 1-butanethiol, 1-pentanethiol, allyl sulfide, propyl sulfide, and butyl sulfide
   25 m GS-Q column or 25 m Poraplot Q column, 150–230° at 10°/min.

7. *n*-Propyl disulfide and *n*-propyl trisulfide
   25 m DB-FFAP column at 100°.

8.

$$
\begin{array}{ccc}
\text{N}\!-\!\!-\!\!\text{C}\!-\!\text{Cl} & \text{N}\!-\!\!-\!\!\text{C}\!-\!\text{Cl} & \text{N}\!-\!\!-\!\!\text{C}\!-\!\text{Cl} \\
\underset{\text{Cl}}{\text{C}}\diagdown_{\text{S}}\diagup\underset{}{\text{C}}\!-\!\text{Cl}, & \underset{\text{Cl}}{\text{C}}\diagdown_{\text{S}}\diagup\underset{}{\text{C}}\!-\!\text{Br, and} & \underset{\text{Br}}{\text{C}}\diagdown_{\text{S}}\diagup\underset{}{\text{C}}\!-\!\text{Br}
\end{array}
$$

   25 m DB-210 column at 120°.

9. Phenanthrylene sulfide and pyrene
   30 m DB-210 column, 150–225° at 5°/min.

B. Packed Columns

1. Air, carbon dioxide, hydrogen sulfide, carbonyl sulfide, propane, and sulfur dioxide
   2 m Chromosorb 101 column at 75°.

2. Methane, ethene, ethane, butene, acetone, furan, dimethylsulfide, isoprene, and methylfuran
   3 m Porapak QS column, 60–200° at 8°/min.

II. Mass Spectra of Sulfur Compounds

A. Aliphatic thiols (mercaptans)

1. General formula: RSH

2. Molecular ion: The presence of sulfur can be detected by the ^{34}S isotope (4.4%) and the large mass defect of sulfur in accurate mass measurements. In primary aliphatic thiols, the molecular ion intensities range from 5–100% of the base peak.

3. Fragment ions: Loss of 34 (H_2S) Daltons from the molecular ions of primary thiols is characteristic. In secondary and tertiary thiols, 33 Daltons rather than 34 Daltons are lost from their molecular ions. If m/z 47 and 61 are reasonably abundant ions, and m/z 61 is approximately 50% of m/z 47, then the mass spec-

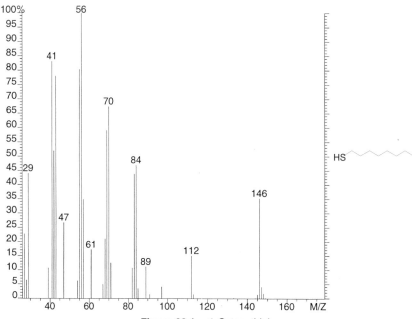

Figure 33.1. 1-Octanethiol.

trum represents a primary thiol (see Figure 33.1). If m/z 61 is much less than 50% of m/z 47, then the mass spectrum represents a secondary or tertiary thiol. Also, remember that saturated sulfur compounds generally fragment beta to the sulfur atom to lose the largest alkyl group.

B. Thioethers (Sulfides)

1. General formula: R-S-R′

2. Molecular ion: Molecular ions are usually reasonably abundant. Cyclic thioethers give abundant molecular ions and an abundant fragment ion due to double β-cleavage (e.g., m/z 60 [CH$_2$-S-CH$_2$]$^+$).

3. Fragmentation: The thioethers can be distinguished from the primary, secondary, and tertiary thiols by the absence of the losses of either 33 or 34 Daltons from their molecular ions. Fragmentation also occurs beta to the sulfur atom with dominant loss of the larger alkyl group. Fragmentation can occur on either side of the sulfur atom with the rearrangement of a hydrogen atom.

C. Aromatic Thiols

1. General formula: ArSH

2. Molecular ion: Aromatic thiols show intense molecular ions (see Figure 33.2).

3. Fragmentation: Characteristic losses from the molecular ion include: 26, 33, and 44 (C=S) Daltons. Ions characteristic of the phenyl group are also observed.

D. Thioesters

1. General formula: RC(O)SR'

2. Molecular ion: The molecular ion of thioesters is more intense than that of the corresponding oxygenated esters.

3. Fragmentation: R^+ and RCO^+ are frequently the most abundant ions observed. $RCOS^+$, SR'^+, and R'^+ are usually observed.

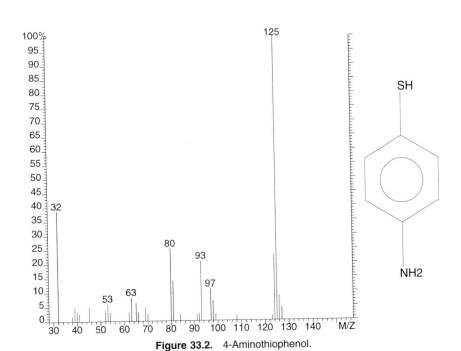

Figure 33.2.   4-Aminothiophenol.

E. Isothiocyanates

1. General formula: RNCS

2. Molecular ion: Molecular ions are observed.

3. Fragmentation: A characteristic fragment ion is observed at
   m/z 72 ($CH_2=N=C=S$). For R>n-C_5, m/z 115 is prominent.

Part II References

Amino Acids

Bierman, C. J. *J. Chromatogr.*, *357*, 330, 1986.
Early, R. J., Thompson, J. R., Sedgwich, G. W., Kelly, J. M., and Christopherson, R. J. *J. Chromatogr.*, *416*, 15, 1987.
Gelpi, E., Koenig, W. A., Gilbert, J., and Oro, J. *J. Chromatogr.*, *7*, 604, 1969.
Kitson, F. G., and Larsen, B. S. In C. N. McEwen and B. S. Larsen, Eds., *Mass Spectrometry of Biological Materials.* New York: Marcel Dekker, 1990.
Marquez, C. D., Weintraub, S. T., and Smith, P. C. *J. Chromatogr. B: Bio. Med. Appl.*, *658*, 213, 1994.
Zhu, P., Yu, Z., Wan, G., Sha, H., Su, K., Yu, B., and Zhao, G. *Org. Mass Spectrom.*, *26*, 613, 1991.

Bile Alcohols

Cohen, B. I., Kuramoto, T., Rothschild, M. A., and Mosbach, E. H. *J. Biol. Chem.*, *251*, 2709, 1976.

Cannabinoids

Harvey, D. H. Chapter 6. In D. M. Desiderio, Ed., *Mass Spectrometry Clinical and Biomedical Applications* (Vol. 1). New York: Plenum Press, 1992.

Carbohydrates

Pitkanen, E., and Kanninen, T. *Biol. Mass Spectrom.*, *23*, 590, 1994.
Previs, S. F., Ciraolo, S. T., Fernandez, C. A., Beylot, M., Agarwal, K. C., Soloviev, M. V., and Brunengraber, H. *Anal. Biochem.*, *218*, 192, 1994.

Carboxylic Acids

Beihoffer, J., and Ferguson, C. *J. Chromatogr. Sci.*, *32*, 102, 1994.
Kawamura, K. *Anal. Chem.*, *65*, 3505, 1993.

Carnitines

Millington, D. S., and Chace, D. H. Chapter 8. In D. M. Desiderio, Ed., *Mass Spectrometry Clinical and Biomedical Applications* (Vol. 1). New York: Plenum Press, 1992.
Van Bocxlaer, J. F., Claeys, M., Van den Heuvel, H., and De Leenheer, A. P. *J. Mass Spectrom., 30,* 69, 1995.

Drugs and Metabolites

Gudzinowicz, B. J. and Gudzinowicz, M. J. *Analysis of Drugs and Metabolites by Gas Chromatography Mass Spectrometry* (Vols. 1–5). New York: Marcel Dekker, 1977.
Jaeger, H., Ed. *Capillary Gas Chromatography–Mass Spectrometry in Medicine and Pharmacology.* Heidelberg, Germany: Heutig, 1987.
LaDu, B. N., Mandel, A. G., and Way, E. L. *Fundamentals of Drug Metabolism and Drug Disposition.* Baltimore: Williams & Wilkins, 1972.
Lui, F., Hu, X. Y., and Luo, Y. *J. Chromatogr. B-Bio. Med. Appl., 658,* 375, 1994.
Pfleger, K., Maurer, H., and Weber, A. *Mass Spectral and GC Data of Drugs, Poisons, Pesticides, Pollutants, and Their Metabolites* (Vol. 1). Weinheim: VCH Publishers, 1992.
Schanzer, W., and Donike, M. *Biol. Mass Spectrom., 1,* 3, 1992.
Sunshine, I, and Caplis, M. *CRC Handbook of Mass Spectra of Drugs.* Boca Raton, FL: CRC Press, 1981.

Dyes

Straub, R. F., Voyksner, R. D., and Keever, J. T. *Anal. Chem., 65,* 2131, 1993.

Environmental

American Chemical Society. *EPA Methods* (Iris, Air Toxics, and Pesticide Methods) (1155 16th Street, N.W., Washington, DC)
Clement, R. E., and Tosine, H. M. *Mass Spectrom. Rev., 7,* 593, 1988.
Hites, R. A. *Mass Spectra of Environmental Contaminants.* Boca Raton, FL: CRC Press, 1992.

Essential Oils

Masada, Y. *Analysis of Essential Oils by GC and MS.* New York: John Wiley, 1976.

Fluorinated Compounds

Simmonds, P. G., O'Doherty, S., Nickless, G., Sturrock, G. A., Swaby, R., Knight, P., Richetts, J., Woffendin, G., and Smith, R. *Anal. Chem., 67,* 717, 1995.

Gasoline Additives

Kania, H., Inouye, V., Coo, R., Chow, R., Yazawa, L., and Maka, J. *Anal. Chem., 66,* 924, 1994.

Schutjes, C. P. M., Vermeer, E. A., Rijks, J. A., and Cramers, C. A. *J. Chromatogr., 253,* 1, 1982.

Lipids

Adams, J., and Qinghong, A. *Mass Spectrom. Rev., 12,* 51, 1993.

Christie, W. W. *Gas Chromatography and Lipids: A Practical Guide.* Alloway: The Oily Press, 1989.

Murphy, R. C. *Mass Spectrometry of Lipids.* New York: Plenum Press, 1993.

Shibamoto, T., Ed. *Lipid Chromatographic Analysis.* New York: Marcel Dekker, 1993.

Microbiology

Odham, G., Larsson, L., and Mardh, P. A., Eds. *Gas Chromatography/Mass Spectrometry: Applications in Microbiology.* New York: Plenum Press, 1984.

Nucleosides

Langridge, J. I., McClure, T. D., El-Shskawi, S., Fielding, A., Schram, K. H., and Newton, R. P. *Rapid Commun. Mass Spectrom., 7,* 427, 1993.

Teixeira, A. J. R., Van de Werken, G., Stavenuiter, J. F. C., De Jong, A. P. J. M., Westra, J. G., and Van der Greef, J. *Biol. Mass Spectrom., 9,* 441, 1992.

PCBs and Other Polyhalogenated Compounds

Ballschmiter, K., Bacher, R., and Mennel, A. *J. High Res. Chromatogr., 15,* 260, 1992.

Samsonnov, D. P., Keryukhin, V. P., and Zhiryukhina, N. P. *J. Anal. Chem., Engl. Tr., 49,* 783, 1994.

Sutton, M. L., Frenkel, M., and Marsh, K. N. *Spectral Data for PCBs.* Boca Raton, FL: CRC Press, 1994.

Pesticides

Cairns, T., and Baldwin, R. A. *Anal. Chem., 67,* 552A, 1995.

Eisert, R., and Levsen, K. *J. Am. Soc. Mass Spectrom., 6,* 1119, 1995.

Yasin, M., Baugh, P. J., Hancock, P., Bonwick, G. A., Davis, D. H., and Armitage, R. *Rapid Commun. Mass Spectrom., 9,* 1411, 1995.

Plasticizers and Polymer Additives

Asamoto, B., Young, J. R., and Citerin, R. J. *Anal. Chem., 62,* 61, 1990.
Cortes, H. J., Bell, B. M., Pfeiffer, C. D., and Graham, J. D. *J. Microcol. Separ., 1,* 278, 1989.

Polymer Analysis

Cho, W. J., Choi, C. H., and Ha, C. S. *J. Polym. Sci-A. Polym. Chem., 32,* 2301, 1994.
MacMahon, T., and Chace, M. *J. Am. Soc. Mass Spectrom., 5,* 299, 1994.
Montaudo, G., Puglisi, C., Blazso, M., Kishore, K., and Ganesh, K. *J. Anal. Appl. Pyrol., 29,* 207, 1994.
Ohtani, H., Luo, Y. F., Nakashima, Y., Tsukahara, Y., and Tsuge, S. *Anal. Chem., 66,* 1438, 1994.

Prostaglandins

Knott, I., Raes, M., Dieu, M., Lenoir, G., Burton, M., and Remacle, J. *Anal. Biochem., 210,* 360, 1993.
Waddell, K. A., Blair, I. A., and Wellby, J. *Biomed. Mass Spectrom., 12,* 399, 1985.
Waller, G. R., Ed. *Biochemical Applications of Mass Spectrometry.* New York: Wiley (Interscience), 1972.

Steroids

Pfaffenberger, C. D., Malinak, L. R., and Horning, E. C. *J. Chromatogr., 158,* 313, 1978.

Tobacco Smoke

Dong, M., Schmeltz, I., Jacobs, E., and Hoffman, D. *J. Anal. Toxicol., 2,* 1978.
Zahlsen, K., and Nilsen, O. G. *Pharmacol. Toxicol., 75,* 143, 1994.

Volatile Organic Compounds and Headspace Analysis

D'Agostino, P. A., and Porter, C. J. *Rapid Commun. Mass Spectrom., 6,* 717, 1992.
Manura, J. J. *American Laboratory,* March 1994.
Rodgers, R. S. *American Laboratory,* December 1993.

Part III

Ions for Determining Unknown Structures

M + 1   Possible Precursor Compounds

^{13}C isotope (present at 1.1% for each ^{12}C)
Aliphatic nitriles (usually also have M − 1)
Ethers
Sulfides
Aliphatic amines
Alcohols
Esters

Increasing the sample size or decreasing the repeller voltage may increase the relative abundance of the M + 1 ion. If the sample pressure is very high, dimers also may be produced.

M⁺·   In general, the relative height of the molecular ion peak decreases in the following order.

Aromatics	Amines
Conjugated olefins	Esters
Alicyclics	Ethers
Sulfides	Carboxylic acids
Unbranched hydrocarbons	Branched hydrocarbons
Ketones	Alcohols

M − 1   Possible Precursor Compounds and Functionalities

Dioxolanes
Some amines
Aldehydes*
Some fluorinated compounds
       e.g., $C_6F_5-CH_2OCH_2C\equiv CH$
Acetals
Segmented fluoroalcohols (some even lose H + HF)
       (e.g., $CF_3CF_2CF_2CH_2CH_2OH$†)
Aryl−CH_3 groups
N−CH_3 groups
Aliphatic nitriles (also may have M + 1)‡

*Aldehydes also lose M − 18, M − 28, and M − 44 ions. Aromatic aldehydes lose M − 1 and M − 29 ions.
†Also lose HF.
‡M − 1 is larger than M + 1, especially in *n*-nitriles.

Aromatic isocyanates
Aromatic phenols
Certain butenols
Certain fluorinated amines*
   e.g., $C_8F_{17}CH_2CHICH_2NH_2$
        or
   $CF_3(CF_2)_7CH_2CH_2CH_2NH_2$

M − 2   Possible Precursor Compounds

Polynuclear aromatics (e.g., dihydroxyphenanthrene)
Ethylsilanes (dimers to heptamers)

2   Structural Significance

H_2

"M − 3"   Possible Precursor Compounds

M − 3 ions must be fragments from a higher mass molecular ion (e.g., $M-CH_3$, $M-H_2O$).

"M − 10"   Add 18 to the higher mass and 28 to the next lower mass to determine M. Look for a M−44 peak to see if it is an aldehyde.

"M − 13"   Not from a molecular ion, but could be the loss of a $CH_3\bullet$ (radical) and the loss of a $C_2H_4\bullet$ (radical) or a CO from the molecular ion. The mass difference is 13.

"M − 14"   Possible Precursor Compounds

Not observed from a molecular ion; however, a mixture of two different compounds may be present. The higher mass ion is M−H. The apparent loss of 14 would be the loss of 15 from the molecular ion.

14   Structural Significance

CH_2 in ketene

*May not detect the molecular ion, but may only observe an M−H and an M−F ion.

M − 15   Possible Precursor Compounds

Methyl derivatives
tert-Butyl and isopropyl compounds
Trimethylsilyl derivatives
Acetals
Compounds with NC_2H_5 groups
Compounds with aryl-C_2H_5 groups
Saturated hydrocarbons do not lose CH_3 from a straight-chain compound.

15   Structural Significance

CH_3 as in CH_3F, $CH_3N=NCH_3$, $CH_3OC(O)OCH_3$,
$CH_3OC(O)(CH_2)_4C(O)OCH_3$

$$CH_3OCH=CH_2, \ CH_3OCF=CF_2, \ CH_3O\overset{\overset{O}{\|}}{\underset{\underset{O}{\|}}{S}}OCH_3,$$

$CH_3OC(O)CH_2C(O)OCH_3$

NH

M − 16   Possible Precursor Compounds

Aromatic nitro compounds*
N-oxides
Oximes
Aromatic hydroxylamines
Aromatic amides (loss of NH_2)
Sulfonamides
Sulfoxides
Epoxides
Quinones
Certain diamines (e.g., hexamethylenediamine)

16   Structural Significance

O, NH_2, CH_4

*Nitro compounds show M−O and M−NO and should have a large *m/z* 30.

M-17   Possible Precursor Compounds

Carboxylic acids (loss of OH)
Aromatic compounds with a functional group containing
   oxygen ortho to a group containing hydrogen
Diamino compounds (loss of NH_3)
Simple aromatic acids
Some amino acid esters
Loss of NH_3 (amines having a four-carbon chain or longer)

17   Structural Significance

OH, NH_3

M − 18   Possible Precursor Compounds

Primary straight-chain and aromatic alcohols (also look for
   a large *m/z* 31*)
Alcohol derivatives of saturated cyclic hydrocarbons
Steroid alcohols (e.g., cholesterol)
Straight-chain aldehydes (C_6 and upward)†
Steroid ketones (peak generally <10%)
Carboxylic acids, particularly aromatic acids with a methyl
   group ortho to the carboxyl group
Aliphatic ethers with one alkyl group containing more than
   eight carbons
Loss of CD_3 from deuterated TMS derivatives

18   Structural Significance

A highly characteristic peak for amines (NH_4)
H_2O

M − 19   Possible Precursor Compounds

Fluorocarbons
Alcohols, $M-(H_2O + H)$

19   Structural Significance

Fluorine compounds (*m/zs* 19, 31, 50, and 69)
Rearrangement ion characteristic of hydroxyl compounds

---

*Primary alcohols: M − 18, M − 33, and M − 46.
†Aldehydes observe M − 28 and M − 44 ions.

Glycols
Acetals
Diols

M − 20   Possible Precursor Compounds

Aliphatic alcohols
Alkyl fluorides and fluoroethers*
Fluorosteroids
Fluoroalcohols
Compounds such as $C_8F_{17}CH_2CH_3$ may not show a molecular
ion but M−HF as the highest mass ion.

20   Structural Significance

n-Alkyl fluorides

M − 21   Possible Precursor Compounds

M−(H_2F) from segmented fluoroalcohols

$$\left(\text{e.g.,} \quad C_8F_{17} - \underset{\underset{CH_3}{|}}{C} = CH \, OH \right)$$

M−(H + HF) from: $CF_3CH_2CH_2CF_2CF_2CF_3$
Segmented fluoroalcohols are represented by
   $CF_3CH_2CH_2OH$, $CF_3CF_2CH_2CH_2OH$,
   $CF_3CF_2CF_2CH_2CH_2OH$, etc.

24   Structural Significance

Acetylene

25   Structural Significance

Acetylene (strong ion observed in spectrum)
Maleic acid
Acrolein
Fluoroacetylene

---

*Molecular ions are usually not observed. In CI, the highest mass observed is $[M + H-HF]^+$.

M − 26   Possible Precursor Compounds

Loss of C_2H_2 from thiophene
Aromatic compounds
Pyrrole
Bicyclic compounds

e.g.,

Isocyanides ($R-N\equiv C$)

26   Structural Significance

Aliphatic nitriles and dinitriles
HCN
Acrylonitrile
Propanenitrile
Pyrazines
Pyrroles
Acetylene
Maleic acid
Maleic anhydride
Succinic anhydride
Vinyl group (look for m/z 40 and 54)

M − 27   Possible Precursor Compounds

N-containing heterocyclics (see pyridine, etc.)
Simple aromatic amines
Loss of C_2H_3 by double H rearrangement in ethyl-containing
    phosphites* and phosphonates*
HCN for unsaturated nitriles

---

*Phosphite $(C_2H_5O)_2POH$, phosphonate $(C_2H_5O)_2P(O)H$, ethyl phosphonate $(RO)_2P(O)C_2H_5$.

27   Structural Significance

Aliphatic nitriles and dinitriles
Alkenes and compounds with unsaturated R groups
$CH_2=CH-R$
$CH_2=CHC(O)R$
$CH_2=CHC(O)OR$, etc.
HCN, $CH_2=CH-$

M − 28   Possible Precursor Compounds

Phenols
Diaryl ethers (especially ethyl ethers)
Quinones
Anthraquinones
Aldehydes (look for M − 1, M − 18, M − 28, and M − 44 ions)
Cyclic ketones
Anhydrides
Naphthols
O-containing heterocyclics
M − (HCN + H) (low molecular weight aliphatic nitriles
  and isocyanides)*
Phenylisocyanate (M−CH_2N)
Phenylenediisocyanate
Chlorophenylisocyanate
N-containing heterocyclic compounds
Ethyl esters
Cycloalkanes
Acid fluorides (e.g., benzoyl fluoride)

28   Structural Significance

Lactones
Ethyleneimines
Alkyl amines
Saturated hydrocarbons
Dialkyl aromatic hydrocarbons
Dioxanes

*Isocyanides lose HCN + H_2CN more readily, and this may help differentiate cyanides from isocyanides.

Cyclobutane
Butyrolactone
Butyraldehyde
Succinic acid
1-Tetralone
Carbonyls

M − 29   Possible Precursor Compounds

Aromatic aldehydes (should observe M − H as well as M − CHO)
Simple phenols
Quinones
Ethyl groups or alicyclic compounds
Cyclic ketones and cyclic amino ketones, such as pyrilidone, piperidone, and caprilactam
Phenols
Naphthols
Polyhydroxybenzenes
Diaryl ethers
Aliphatic nitriles
Purines $[M-(CH_2=NH)]$ and imines
Propanals
Indicates an alkyl group (may not be ethyl)
Loss of CHO from compounds such as $CH_3C(O)O-CH_2OC(O)CH_3$

29   Structural Significance

Alkanes or compounds with an alkyl groups (m/z 43, 57, 71, and so forth may also be present)*
Aldehydes
Propionates
Cyclic polyethers
α-Amino acids
Hydroperoxides
C_2H_5, CHO

*For branched alkanes such as *t*-butyl m/z 43 is the dominant ion.

M − 30 Possible Precursor Compounds

Aromatic nitro compounds
Loss of CH_2O from simple aromatic ethers
Lactones*

Loss of CH_2O from [structure]   (usually do not see the molecular ion)

(see dioxane, dioxolanes, and epichlorohydrin)
Morpholine

30 Structural Significance

Cyclic amines unsubstituted on the nitrogen
Primary amines†
Nitro compounds and aliphatic nitrites (should observe M −
    16 and M − 30)
Secondary amides
Formamides
Nitrosamines
Ureas
Caprolactam
NO (aliphatic and aromatic nitro compounds)
N_2H_2, CH_2O, CH_4N, CH_2NH_2, C_2H_6, etc.

Note: In the absence of rearrangement, fragment ions containing an odd number of nitrogen atoms will be of even mass.

M − 31 Possible Precursor Compounds

Methyl esters—the simultaneous presence of peaks at *m/z*
    74 and 87 and M-31 indicates a methyl ester

*Produces M + H using isobutane CI.
†Amines show a powerful tendency to fragment beta to the nitrogen atom. This mass is not characteristic only of primary compounds, because the ion is formed from other amines by rearrangement.

Methoxy derivatives, including methoximes
Primary aliphatic alcohols and glycols
Loss of SiH_3 from compounds like $H(SiH_3C_6H_5)_3H$

31   Structural Significance*

A peak occurs at m/z 31 in almost all alcohols and ethers, as
  well as in some ketones. m/z 31 in the absence of fluorine
  indicates oxygenated compounds.
Fluorocarbons (with m/z 19, 31, 50, and 69)
Ethers and alcohols (m/z 31, 45, and 59)†
Primary straight-chain alcohols
Primary alcohols bonded at the γ-carbon
Formates
Aliphatic carboxylic acids
Alkyl ethyl ethers
Cyclic polyethers
Phosphorous compounds
Rearrangement peak in dioxanes

M − 32   Possible Precursor Compounds

o-Methyl benzoates
Loss of methanol
Ortho substituent of methyl esters of aromatic acids. Loss of
  31 also should be present.
M−(CH_3OH) from methyl esters and ethers
M−(O + NH_2) from sulfonamides
Loss of sulfur from thiophenols and disulfides

32   Structural Significance

Hydrazine
Methanol
Fluoroethylene
Deutero compounds
Oxygen

---

*m/z 31 in the absence of fluorine indicates oxygenated compounds.
†If it is an alcohol, you may have to add 18 or 46 to the highest m/z observed to identify the
molecular weight.

Methyl formate
Carbonyl sulfide

M − 33   Possible Precursor Compounds

Alcohol derivatives of cyclic hydrocarbons
Thio compounds
Thiocyanates
Short-chain unbranched primary alcohols
$M-(H_2O + CH_3)$ from hydroxy steroids
Loss of $-SH$

33   Structural Significance

Glycols
Diols
Alcohols
Acetals
CH_2F

M − 34   Possible Precursor Compounds

Mercaptans (usually aliphatic)

34   Structural Significance

CH_3F, H_2S, PH_3 (see triethyl phosphine, etc.)

M − 35   Possible Precursor Compounds

Chloro compounds
$M - (H_2O + OH)$ from certain dihydroxy and polyhydroxy compounds
$M - (CH_3 + HF)$ from compounds such as

$$\overset{\displaystyle CF_3}{\underset{\displaystyle (CH_3)_3SiOCH_2CF_2OCF}{|}} —$$

35   Structural Significance

Thioethers
Sulfides
Chloro compounds

M − 36 Possible Precursor Compounds

n-Alkyl chlorides (Some alkyl chlorides lose HCl and appear to be butenes [e.g., 1,2,3,4-tetrachlorobutane appears to be trichlorobutene].)
M − (H_2O+H_2O) from certain polyhydroxy compounds

36 Structural Significance

HCl

M − 37 Possible Precursor Compounds

M − ^{37}Cl from alkyl chlorides
(H_2O + F) from certain fluoroalcohols

37 Structural Significance

HC—C≡C—

M − 38 Possible Precursor Compounds

M − 2F from fluorocarbons

38 Structural Significance

C_2N, $-CH_2C{\equiv}C-$
Diallyl sulfide
Malononitrile
Dicyanoacetylene
Furan

M − 39 Possible Precursor Compounds

(HF + F) (e.g., fluoroalcohols)

$$CH_3OCF_2CF_2\underset{\underset{OH}{|}}{\overset{\overset{H}{|}}{C}}CF_2CF_2OCH_3$$

39 Structural Significance

Furans
Alkenes

Dienes
Cyclic alkenes
Acetylenes
Aromatic compounds, particularly di- and tetrasubstituted
  compounds (*m/z* 39, 50, 51, 52, 63, and 65)
C_3H_3, $CH_3C\equiv C-$, $-CH=C=CH_2$, $-CH_2C\equiv CH$

M − 40   Possible Precursor Compounds

Aliphatic dinitriles (loss of CH_2CN)*
Aromatic compounds
Cyclic carbonate compounds
Segmented fluoroalcohols (i.e., loss of H_2F_2)

$$C_8F_{17} - \underset{\underset{CH_3}{|}}{C} = CHOH$$

[M-(HF+HF)]

40   Structural Significance

Butanol
Dinitriles (may have *m/z* 41, 54, and 68)
C_3H_4, CN_2, C_2O, $CH_3C\equiv CH$, $CH_2=C=CH_2$, C_2H_2N,
  $-CH_2C\equiv N$

M − 41   Possible Precursor Compounds

Nitriles
Suggests a propyl ester

N-influenced fragmentation, e.g.,

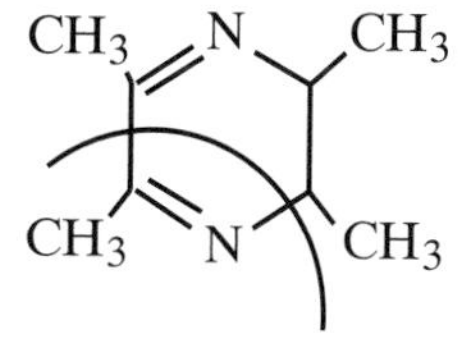

---

*The highest mass may be M − 28 and/or M − 40. (Usually the molecular ion is not observed.)
Large *m/zs* 41, 55, and 54 peaks also should be present.

41

Structural Significance

Nitriles and dinitriles (*m/z* 54 also may be present)*
Esters of aliphatic dibasic carboxylic acids
Thioethers
Isothiocyanates
Primary aliphatic alcohols
Alkenes and compounds with an alkenyl group
Cyclohexyl (*m/z* 41, 55, 67, 81, and 82)
Methacrylates (also look for *m/z* 69)†

$$C_3H_5, C_2HO, C_2H_3N, CH_3CN, -CH_2CH=CH_2,$$
$$\overset{\displaystyle CH_3}{\underset{\displaystyle |}{CH_3CH=CH-, -C=CH_2}}$$

m/z 69, 41, and 86 suggest segmented fluoromethacrylates‡

M − 42

Possible Precursor Compounds

Acetates (loss of ketene, see if *m/z* 43 is also present)
N-acetylated compounds (acetamides)
Enol acetates
Simple purines (loss of cyanamide)
2-Tetralone
Loss of ketene from bicyclic structures (e.g., for camphor)

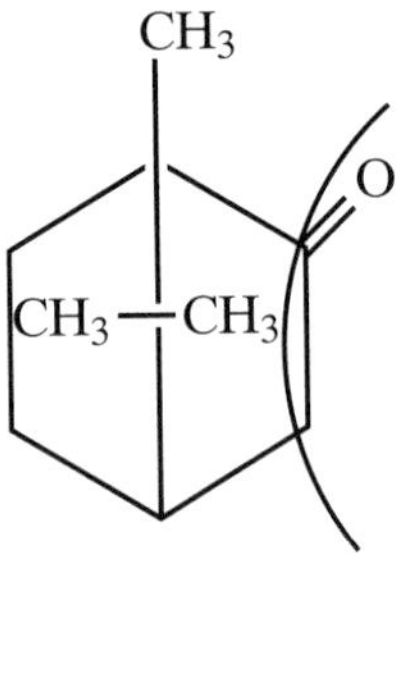

*CH₃CN exists in lower aliphatic nitriles. Higher nitriles have $(CH_2)_n$ CN (e.g., *m/z* 54, 68, etc.)
†May have to add 86 Daltons to last peak observed to deduce the molecular weight.
‡$C_nF_{2n+1}CH_2CH_2O\overset{\displaystyle O}{\overset{\displaystyle ||}{C}}-\overset{\displaystyle CH_3}{\underset{\displaystyle |}{C}}=CH_2$

42   Structural Significance

Cycloalkanes
Alkenes ($<C_6$); peaks at *m/z* 42 and 56 suggest alkenes
Ethyleneimines ($H_2C=C=NH$)
Pteridines
Cyclic ketones (saturated)
THF
Butanediol
Simple purines
C_3H_6, C_2H_2O, CNO, CH_2N_2, C_2H_4N, $CH_3N=CH-$

$$\underset{CH_2}{\overset{CH_2}{|}} \!\!\! \diagdown N, \ CH_3CH=N-, \ -\overset{\overset{\textstyle H}{\overset{\textstyle N}{\diagup\diagdown}}}{CH}\!\!-\!\!CH_2, \ CH_2=N=CH_2$$

$(CH_3)_2N-$ also should observe *m/z* 44

M − 43   Possible Precursor Compounds

Uracils: M-(C(O)NH)
Cyclic amides
Propyl derivatives (particularly isopropyl)
Cyclic peptides
Dioxopiperazines
Tertiary amides
Aliphatic nitriles ($>C_6$)
Common loss from cyclohexane rings
Terpenes

43   Structural Significance

Alkanes (*m/z* 43, 57, 71, 85, etc., characterize the spectra
    of alkanes)
Compounds with an alkyl group
Acetates (also observe *m/z* 61)
Alditol acetates (Check for the presence of *m/z* 145, 217,
    289, etc.)

Methyl ketones (When looking for larger than methylethyl ketone, look for *m/z* 58 as well.)
Cyclic ethers
Vinyl alkyl ethers
N-alkylacetamides
Isothiocyanates
Aliphatic nitriles
With *m/z* 87 suggests triethyleneglycol diacetate
C_3H_7, CH_3C-, $(CH_3)_2CH$, $CH_2=CHO$, C_2H_5N, CH_3N_2, $C(O)NH$, $C_3H_7C(O)OR$, $C_3H_7C(O)NR_2$, $C_3H_7C(O)R$, C_3H_7SR, C_3H_7OR, $C_3H_7OC(O)R$, C_3H_7X, $C_3H_7NO_2$

M − 44   Possible Precursor Compounds

Aliphatic aldehydes (Look for M − 1, M − 18, M − 28, and M − 44 ions.)
Compounds with $(CH_3)_2N-$
Aromatic amides
Anhydrides (e.g., phthalic acid anhydride)

44   Structural Significance

Aliphatic aldehydes unbranched on the α-carbon (is the base peak for C_4-C_7) (Look for M − 1, M − 18, M − 28, and M − 44 ions.)
Primary amines with α-methyl groups
Secondary amines unbranched on the α-carbon
Tertiary amides

$$\underset{R_1C-N}{\overset{\overset{\displaystyle O}{\|}}{}}\diagdown\begin{matrix}R_2{}^* \\ \\ R_3\end{matrix}$$

Vinyl alkyl ethers
Caprolactam derivatives

*Will produce $CH_3C(O)\text{-}N\diagdown\begin{matrix}R_2\\ \\ R_3\end{matrix}$ by a McLafferty rearrangement provided R1 contains a hydrogen on carbon 4.

Cyclic alcohols
Piperazines (C_2H_6N)
CO_2, N_2O, $CF\equiv CH$, CH_3CHO, C_2FH, C_3H_8, $OCNH_2$,
C_2H_4O, CH_2NHCH_3, $(CH_3)_2N$, CH_2NO, $CH_2CHO + H$,
CH_3CHNH_2, CH_4N_2, $CH=NOH$, $C_2H_5NHCH_2CH_2NH_2$

M − 45 Possible Precursor Compounds

Ethoxy derivatives
Ethyl esters
Carboxylic acids (may lose 17, 18, and 45)
Cyanoacetic acid, 8-cyano-1-octanoic acid

45 Structural Significance

Aliphatic carboxylic acids
2-alcohols (e.g., *sec*-butyl alcohol, *m/z* 45, 59, and 31)*
Propylene glycol
Di- and triethylene glycols
Isopropyl and *sec*-butyl ethers
Butyrates
Methylalkyl ethers unbranched on the α-carbon
$-C(O)-OH$, $CH_2=CF-$, C_2H_5O, CH_3OCH_2, CH_2CH_2OH,
C_2H_2F, CH_3NO, C_2H_7N, CH_3SiH_2, $CH_2=P^+$, CH_3CHOH,
$CHS\dagger$
Oxygenated compounds can be disregarded if peaks are absent at *m/z* 31, 45, 59, etc.

M − 46 Possible Precursor Compounds

Nitro compounds (NO_2)
Loss of C_2H_5OH from ethyl esters
Aromatic acids with methyl groups ortho to the carboxyl groups
Long-chain unbranched primary alcohols‡
Loss of C_2H_3F
Straight-chain high molecular weight primary alcohols
Loss of water plus C_2H_4

*$-CHOH$ (2-alcohols)
|
CH_3

†Is characteristic of certain sulfur compounds such as thiophenes.
‡Add 48 daltons to the last *m/z* of reasonable intensity to deduce the molecular weight.

Possibly loss of formic acid [e.g., $C_6H_5CH_2CH_2C(O)OH$]
Cyano acids

46   Structural Significance

Nitrates
$CH_2=CHF$, CH_3CH_2OH, $HC(O)OH$
CH_3OCH_3, NO_2, C_2FH_3, CH_3NH NH_2

M − 47   Possible Precursor Compounds

Alkylnitro compounds
Acid fluorides
Sulfur compounds (loss of CH_3S)
Loss of CH_3OH + CH_3

47   Structural Significance

Thiols (m/z 61 and 89 also suggest sulfur-containing compounds)
Acid fluorides
Thioethers (sulfides)
Fluorosilanes such as $FC_6H_4SiH_3$
Acetals (generally containing the ethoxy group)
Fluoroethane
C, O, and F compounds (peak may be small)
Formates (small peak)
CH_3S, $-C(O)F$, SiF, PO
CH_3CHF, CH_2SH, CH_3PH, CH_2CH_2F

M − 48   Possible Precursor Compounds

Methyl thioethers
(CO + HF) from pentafluorophenol
(CHO + F) from fluoroaldehydes
Aromatic sulfoxides

48   Structural Significance

Methyl thioethers
CH_3SH from mercaptans
SO, CH_3SH, $-CHCl$, C_2H_5F, CH_3S +H, C_4
Tetraborane

M − 49   Possible Precursor Compounds

Chlorinated compounds (loss of CH_2Cl) (e.g., β-chloroisopropylbenzene)

49   Structural Significance

Halogenated compounds containing CH_2Cl
Methyl thioethers
CH_2Cl, $-C\equiv C-C\equiv CH$

M − 50   Possible Precursor Compounds

Fluorocarbons (loss of CF_2)
Methyl esters of unsaturated acids (loss of $CH_3OH + H_2O$)
Methyl esters of straight-chain hydroxycarboxylic acids except 2-hydroxy acids

50   Structural Significance

CF compounds
Compound containing phenyl or pyridyl groups
Chloromethyl derivatives
CF_2

M − 51   Possible Precursor Compounds

Loss of CHF_2
$HC\equiv CCN$ from α,β-unsaturated nitriles

51   Structural Significance

Acetylenes
Compounds containing phenyl or pyridyl groups
Compounds containing CHF_2
Small m/z 51 peak indicates compounds containing C, H, and F
CHF_2, C_4H_3 (aromatics), SF

M − 52   Possible Precursor Compounds

52   Structural Significance

Butadienes
Acetylenes

Compounds containing phenyl or pyridyl groups
Nitrogen trifluoride
C_2N_2, $-CH=CHCN$, $CH_2=CH-C\equiv CH$, C_2H_4, NF_2
Chromium
Cyanogen, C_2N_2

M − 53   Possible Precursor Compounds

53   Structural Significance

Furans
Cyclobutenes
Dienes
Acetylenes
Pyrazine
Aminophenol
Acrylonitrile
Chloroprene
Hydroquinone
C_3H_3N, C_4H_5, C_2HN_2, C_3HO, NF_2H, etc.
The mass spectra of pyrrole derivatives usually contain promi-
nent ions at *m/z* 53, and 80, and sometimes the *m/z* 67
rearrangement ion.

M − 54   Possible Precursor Compounds

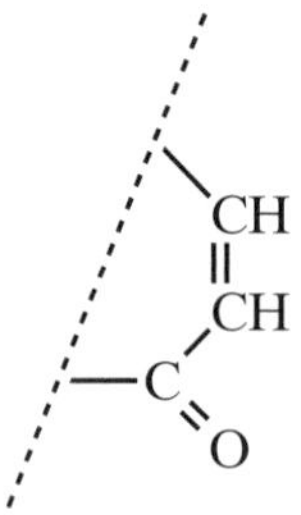

-CHCN   (from branched dinitriles)
|
CH_3

54   Structural Significance

Unsaturated cyclic hydrocarbons (e.g., vinyl-cyclohexene)
Butadienes
Dinitriles (C_2H_4CN) (Dinitriles lose 40 and 28 Daltons and
generally give *m/z* 41, 55, and 54 ions.)
Acetylenes
Cyclohexanes
Nitriles (*m/z* 41 also may be present)
Maleic acid
Maleic anhydride
Quinone
$-CH_2CH_2CN$, $C(O)CN$, C_3H_2O, C_4H_6, $C_2H_2N_2$

M − 55   Possible Precursor Compounds

A loss of 55 is possibly the loss of C_4H_7 from esters (double
hydrogen rearrangement). The loss suggests a butyl or isobutyl group, especially when *m/z* 56 is also present.
Loss of (CO+HCN) from aromatic isocyanate
Loss of $CH_2=CH-C(O)$

55   Structural Significance

Cyclic ketones
Cycloalkanes
Indicates a cyclohexyl ring (*m/z* 55, 83, and 41)
Esters of aliphatic dibasic carboxylic acids
Aliphatic nitriles and dinitriles (see *m/z* 54)
Thioethers
Alkenes
Primary aliphatic alcohols
Olefins have fragment ions at *m/z* 55, 69, 83, etc.
Cyclohexanones
$(CH_3)_2C=CH$, $CH_2=CH-C(O)-$
Acrylates (*m/z* 55 and 99 suggest glycol diacrylates)

Base peak	Next most intense peaks				Compound	Highest m/z peak >1%
55	27	82	67	54	1,6-Hexanediol diacrylate (MW = 226)	99
55	27	82	81	56	Pentaerythnitol triacrylate (MW = 298)	196, 225
55	54	27	71	85	1,4-Butylene glycol diacrylate (MW = 198)	
55	54	27	71	85	1,4 Butylene glycol diacrylate (MW = 198)	
55	54	71	126	85	1,4-Butylene glycol diacrylate (MW = 198)	126
55	56	70	43	69	n-Octyl acrylate (m/z 84 and 83)	112
55	56	70	73	98	n-Heptyl acrylate (MW = 120)	98
55	56	73	27	41	Butyl acrylate (MW = 128)	99, 113
55	56	73	84	43	n-Hexyl acrylate (MW = 156)	
55	59	72			2-Ethoxyethyl acrylate (MW = 144)	
55	59	72	43	45	C₂H₅OCH₂CH₂OCH=CH₂ (MW = 144)	99, 100, 101
55	67	82	73		Cyclohexyl acrylate (MW 154)	
55	81	82	126	127	Pentaerythritol triacrylate (MW = 298)	196, 225
55	82	54	67	27	1,6-Hexanediol diacrylate (MW = 226)	
55	84	43	69	56	2-Ethylbutyl acrylate (MW = 156)	
55	85	45			2-Methoxyethyl acrylate (MW = 130)	
55	85				Methyl acrylate (MW = 86)	86
55	86	73	85		Ethylene glycol diacrylate (MW = 170)	
55	91	117	162		Benzyl acrylate (MW = 162)	162
55	98	127	68	70	Ethylene glycol diacrylate (MW = 170)	127
55	99	27	100	56	Diethylene glycol diacrylate (MW = 214)	129
55	99	77			2-Phenylethyl acrylate (MW = 192)	
55	127	56	27	68	2,2-Dimethyl propane diacrylate (MW = 212)	140, 152

M − 56   Possible Precursor Compounds

Aromatic diisocyanates
Quinones
Ketals of cyclohexanone
Anthraquinones
C₄H₈ from carbonyl compounds
Butyl compounds (e.g., C₆H₅OC₄H₉)

56   Structural Significance

Butyl esters
Cyclohexylamines
Isocyanates (n-alkyl). Peaks associated with M−CO, M−(H+CO), M−(HCN+CO) are generally observed. Also look for m/z 99.

Cycloalkanes
Some *n*-chain alcohols (butanol, hexanol, heptanol, octanol, nonanol, etc.)
Aliphatic nitriles (see *m/z* 54)
$-CH_2-N=C=O$, C_4H_8, C_3H_6N, $(CH_3)_2N=CH_2$, $CH_3N=CHCH_2-$

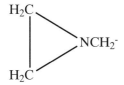

M − 57   Possible Precursor Compounds

Isocyanates $(M-CH_3N=C=O)$
Loss of F_3 in perfluorotributylamine
C_4H_9 from TBDMS derivatives

57   Structural Significance

Compounds containing alkyl groups, particularly tertiary butyl group
Ethyl ketones
Propionates
Aliphatic nitriles
Isobutylene trimers and tetramers (also observe *m/z* 41 and 97)
Alkanes (*m/z* 29, 43, 57, 71, etc. are also observed)
C_4H_9; $C_2H_5CHCH_3$, $(CH_3)_3C$, etc.
C_3H_5O; $C_2H_5C(O)-$, $CH_3C(O)CH_2-$, etc.
$C_2H_5N=N$

M − 58   Possible Precursor Compounds

Aliphatic methyl ketones $(CH_3C(O)CH_3)$
Simple aromatic nitro compounds $(NO+CO)$
Straight-chain mercaptans
Thiocyanates and isothiocyanates

58   Structural Significance

Methyl alkyl ketones unbranched at the α-carbon (43+58)

$$\overset{.+}{\underset{|}{O}}H$$

$(CH_2=C-R)$

Rearrangement ion of 2-methylaldehydes
n-Propyl-, butyl-, amyl-, and hexylketones
Primary amines with an 2-ethyl group
Tertiary amines with at least one *N*-methyl group
Thiocyanates
Isothiocyanates

$$\overset{O}{\underset{||}{HCNHCH_2}}\text{-,}\quad \overset{CH_3}{\underset{|}{-C}}\text{=NOH,}\quad \overset{NH}{\underset{||}{-C}}\text{-OCH}_3,$$

$(CH_3)_2CNH_2, (CH_3)_3CH, C_3H_8N$

$(CH_3)_2NCH_2-$, (sometimes a *m/z* 42 peak also is present), $C_2H_5NHCH_2-$
$CH_3C(O)NH-$ (generally observe a *m/z* 60 rearrangement ion)
$CH_3C(O)CH_2+H$, C_3H_6O, $C_2H_2O_2$, C_2H_4NO, $C_2H_6N_2$, C_3H_8N, $-C(O)NHCH_3$, CNO_2, C_3H_3F, CH_2N_2O, C_4H_{10}, $HC(O)NHCH_2-$
Peaks at *m/z* 58, 72, and 86 suggest that a C=O group is present (also look for M-CO).

M − 59   Possible Precursor Compounds

Loss of $C(O)OCH_3$ from methyl esters
Loss of OC_3H_7 from propyl esters
Methyl esters of 2-hydroxycarboxylic acids
Loss of $CH_3C(O)O$ from sugar acetates

59   Structural Significance

Tertiary aliphatic alcohols
Methyl esters of carboxylic acids
Esters of *n*-chain carboxylic acids
Silanes

Ethers ($C_2H_5OCH_2-$)
m/z 59, 45, 31, and 103 suggests di- or tri-propylene glycol
Primary straight chain amides (greater than propionamide)

$$\overset{\overset{\text{OH}}{|}}{}$$
Amides ($CH_2=C\text{-}NH_2$)   (Look for *m/z* 59, 44, and 72)

(CH_3)$_2$COH (tertiary alcohols)
Largest peak in the mass spectrum of propylene glycol ethers
  $CH_3OCH_2CH_2-$ (diethylene glycol dimethyl ether)
$CH_3CHOHCH_2-$, $C_2H_5OCH_2-$, (CH_3)$_2$COH
$CH_3OC(O)-$, CH_3CHCH_2OH, C_3FH_4
Peaks at *m/z* 31, 45, and 59 reveal oxygen in alcohols, ethers, and ketones.

M − 60   Possible Precursor Compounds

Acetates
Methyl esters of short-chain dicarboxylic acids
Loss of $-OCH_2CH_2NH_2$
$-C_6H_{10}(CO_2CH_3)_2$
O-methyl toluates

60   Structural Significance

A characteristic rearrangement ion of monobasic carboxylic acids above C_4 (also see *m/z* 73)

Cyano acids (observe M − 46 and large *m/z* 41, 54, and 55 peaks)
Esters of nitrous acids
Sugars
Aliphatic nitriles unbranched at the α-carbon
Ethyl valerate has *m/z* 60 and 73. (These ions are normally associated with carboxylic acids.)

Cyclic sulfides
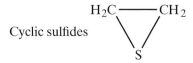

m/z 60 with an *m/z* 58 peak suggests $CH_3C(O)NH+2H$
C_3H_5F, C_3H_8O, CH_6N_3, $-CH_2ONO$, CH_4N_2O, CH_2NO_2,

C_2H_6NO, C_2H_4S, $-NHCH_2CH_2OH$, $ClC\equiv CH$, COS, $C_2H_4O_2$, $C_2H_8N_2$

M − 61   Possible Precursor Compounds

Suggests $CH_3\overset{\bullet}{N}OCH_3$ and CH_3NHOCH_3

$$\text{Loss of } (Si_2H_5) \text{ from} \quad \underset{\underset{H}{|}}{\overset{\overset{Ph}{|}}{H}}Si - \underset{\underset{H}{|}}{\overset{\overset{Ph}{|}}{}}Si - \underset{\underset{H}{|}}{\overset{\overset{Ph}{|}}{}}SiH$$

61   Structural Significance

m/z 61 is a characteristic rearrangement ion in acetates other than methyl acetate (should observe M − 42 as well as a large m/z 43)

Esters of high molecular weight alcohols $(CH_3CO_2H_2)$
Base peak in 1,2-difluorobutanes
Acetals
Primary thiols (CH_2CH_2SH)
Thioethers (CH_3SCH_2-)
N-TFA of *n*-butyl methionine
m/z 61 and 89 suggest sulfur-containing compounds (RSR′)

$$\text{TMS derivatives, e.g.,} \quad -\underset{\underset{H}{|}}{\overset{\overset{CH_3}{|}}{Si}}-OH$$

CH_3CH_2CHF-, $CH_3C(O)C+2H$, CH_3CFCH_3
$-CHOHCH_2OH$, $-CH_2CH_2SH$, CH_3CHSH
CH_3NOCH_3+H

M − 62   Possible Precursor Compounds

Thiols

62   Structural Significance

Ethyl thioethers
$CF_2=C$, $-CHCH_2Cl$, $-CF=CF-$, $PH_2C_2H_5$ (see triethyl-phosphine and tributylphosphine)
CH_3CH_2SH, $(CH_3)_2S$

M − 63   Possible Precursor Compounds

Methyl esters of dibasic carboxylic acids

63   Structural Significance

Aromatic nitro compounds
Acid chlorides $(C(O)Cl)$
Ethyl thioethers
$CF_2=CH-$, $-CH_2CH_2Cl$, CH_3CHCl, $CH_3S(O)$, $SiCl$, C_2F_2H, CH_3POH, $CF_2=CH-$
$CF_3CH_2CH_2$—Structure also provides a large m/z 63 peak.
$C_2H_5OC(O)$ (e.g., diethyl carbonate)
$-CH_2CH_2Cl$, $-OCH_2CH_2Cl$

M − 65   Possible Precursor Compounds

Loss of OCH_2Cl
$-SO_2CH_2-$
Certain sulfones (HSO_2)

65   Structural Significance

Aromatic compounds (C_5H_5)
Aromatic nitro compounds
Aromatic alcohols
Vinyl furans
CH_3CF_2-, $C_2F_2H_3$

M − 66   Possible Precursor Compounds

66   Structural Significance

Unsaturated nitriles
Dicyanobutenes
Methyl pyridines
Ethyl disulfides $(HSSH)$
Acrylonitrile dimers

CFCl, N$_2$F$_2$, C$_2$F$_2$H$_4$, C$_5$H$_6$

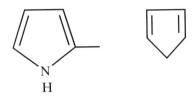

M − 67   Possible Precursor Compounds

67   Structural Significance

Perfluoro acids
Cycloalkyl compounds
Alkadienes (also may have an intense *m/z* 81 ion)
Alkynes

CHClF (in Freon-21), CF$_2$OH, SOF-, C$_5$H$_7$, C$_4$H$_5$N,

CH$_2$=C-C=CH$_2$, *m/z*s 67, 81, 95, 109, etc.
suggest 1-acetylenes

M − 68   Possible Precursor Compounds

Dinitriles

68   Structural Significance

Cyclopentanes
Cyclohexenes
Cyclohexanols
Aliphatic nitriles (*m/z* 41 and 54 also may be present), C$_4$H$_6$N
C$_5$H$_8$ (cyclopentane derivatives), −CH$_2$CH$_2$CH$_2$CN,
    NCCH$_2$C(O)
CH$_2$ClF

M − 69   Possible Precursor Compounds

Fluorocarbons (loss of CF$_3$)

69   Structural Significance

Fluorocarbons
Cycloparaffins (e.g., cyclopentyl)
Esters of aliphatic dibasic carboxylic acids
Aliphatic nitriles
Isothiocyanates ($>C_7$)
Methacrylates (m/z 41 is also present)*
$C(O)C(CH_3)=CH_2$, $CH_3CH=CHC(O)-$, $(CH_3)_2C=CH-$
CH_2-, CF_3, C_5H_9, PF_2, $-CH=CHCH(CH_3)_2$
A series of peaks at m/z 69, 83, 97, 111, etc. (two mass units
less than the corresponding paraffins) suggests olefins
m/z 69, 41, and 86 suggest segmented fluoromethacrylates
m/z 69, 77, 65, and 51 are characteristic of segmented fluoro-
iodides (e.g., $C_8F_{17}CH_2CH_2I$, etc.)

Abundant ions				Compound	MW
69	41	87	39	Ethylene glycol monomethacrylate	130
69	41	113	112	Ethylene glycol dimethacrylate	198
69	113	41	86	Diethylene glycol dimethacrylate	242
69	41	113	45	Triethylene glycol dimethacrylate	286
69	41	55	43	Trimethylolpropane trimethacrylate	338
69	41	54	55	1,6-Hexanediol dimethacrylate	254

M − 70   Possible Precursor Compounds

(CHF_2+F) loss observed in 1-H perfluoro compounds (e.g.,
$CF_3(CF_2)_7H$, $CF(CF_2)_8H$), $-C_4H_8N$

70   Structural Significance

Pyrrolidines

---

*To postulate the molecular ion, add 86 Daltons to the even-mass ion of a doublet in the
high-mass region.

Aliphatic nitriles

Amyl esters

If the mass spectrum has a large peak at m/z 42, check 1-pentanol.

C_4H_8N, $CH_3N=CHCH_2CH-$

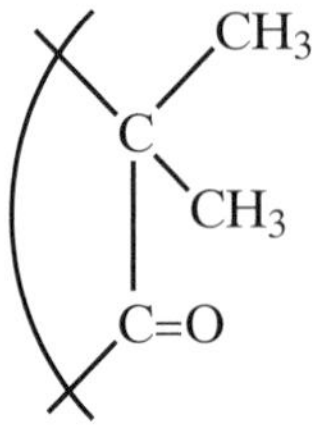

Hexamethyleneimine

Isocyanates ($-CH_2CH_2NCO$)

m/z 70 and 43 suggest a diacetoxybutene.

M − 71   Possible Precursor Compounds

Loss of $(CH_3)_3CCH_2$

$(CH_3)_3CCH_2 \text{-\textsection-} C(CH_3)_2C_6H_4OH$

71   Structural Significance

THF derivatives

Butyrates

Methyl cyclohexanols

Propyl ketones

Terminal aliphatic epoxides, $CH_2\text{-}CHCH_2CH_2\text{-}$ with O bridge

, $CH_3C=CH\text{-}CH_2\text{-}$ (with OH), $CH_2=CHCH_2OCH_2\text{-}$,

$CH_3CH(OH)CH=CH\text{-}$, $C_3H_7C(O)\text{-}$,

$(CH_3)_2CHC(O)$, $\text{-}C(O)OCH=CH_2$,

$$CH_3\diagdown \quad O$$
$$C$$
$$CH_3\diagup \diagdown CH, \qquad CH_2=CH-CH-, \ HOCH_2-C=CH-,$$

with CH_3 and OCH_3 substituents shown on the $CH_2=CH-CH-$ chain.

C_5H_{11}, C_4H_9N, NF_3, $(C_2H_5)_2CH$

M − 72 Possible Precursor Compounds

Loss of $-NHCH_2CH(CH_3)_2$
Acrylates (loss of acrylic acid)
Loss of C_2O_3 from compounds such as cyclohexane dicarboxylic acid anhydride

72 Structural Significance

Amides
Secondary and tertiary amines ($C_3H_{10}N$)
Ethyl ketones
Thiocyanates (CH_2SCN)
Isothiocyanates ($-CH_2-N=C=S$)
Rearrangement ion 2-ethylaldehydes
$CH_3C(O)NHCH_2$, $C_2H_5NHCH_2CH_2-$, $CH_3C(O)CHCH_3+H^*$
$(CH_3)_2CNHCH_3$, $C_3H_7CHNH_2$, $C_2H_5C(O)CH_2+H^*$

$(C_2H_5)_2N$, $(CH_3)_2NC(O)-$, CH_3NCH_2-, $-CHCH_2CH_3$ with $HNCH_3$ substituent

M − 73 Possible Precursor Compounds

Alditol acetates
Loss of $(C(O)C_2H_5)$
Methyl esters of aliphatic dibasic carboxylic acids
$\quad(-CH_2C(O)OCH_3)$
n-Butyl esters (loss of C_4H_9O)†

*By McLafferty rearrangement.
†See dibutyl azelate (MW = 300) and dibutyl oleate (MW = 338).

73   Structural Significance

Sugars
Aliphatic acids (m/z 73 with a peak at m/z 60 suggests
  acids)
Alcohols (C_3H_7CHOH-)
Ethers ($C_3H_7OCH_2-$)
Ethyl esters ($-C(O)OC_2H_5$)
Methyl esters of dibasic carboxylic acids ($CH_2C(O)OCH_3$)

$$CH_2O$$

1,3-Dioxolanes, CH_2 CH-, (m/z) 73 and 45 suggest

dioxolanes
TBDMS and trimethylsilyl derivatives, $(CH_3)_3Si-$

$$OCH_3$$
$$CH_3CH_2CH-, C_2H_5O\overset{\bullet}{C}HCH_3, CH_3C-,$$
$$OCH_3 \qquad CH_3$$
$-CH_2CH_2C(O)OH$

$CH_2OC(O)CH_3$, O   O, C_3H_2Cl
                    C
                    |
                    H

M − 74   Possible Precursor Compounds

Some butyl esters (C_4H_9OH by rearrangement)
(CH_2CHSCH_3) loss from N-TFA, n-butyl methionine

74   Structural Significance

m/z 74 is a characteristic rearrangement ion of methyl esters
of long-chain carboxylic acids in the $C_4–C_{26}$ range. Also look
for a m/z 87 peak.

Monobasic carboxylic acids with an α-methyl group
Esters of aliphatic dibasic carboxylic acids
Cyclic sulfides
Aliphatic nitriles with an α-methyl group

$$\overset{\displaystyle OCH_3}{\underset{\displaystyle |}{}}$$

-CH$_2$NCH$_3$, CH$_3$NHCH$_2$CHOH-, HOC$_2$H$_4$NHCH$_2$-,
CH$_3$CHONO,-CH$_2$NHCH$_2$CH$_2$OH, CH$_2$=C(OH)OCH$_3$,
C$_3$F$_2$,

H$_2$NCHC(O)OH*, $-$N$\overset{\displaystyle CH_3^*}{\underset{\displaystyle CH_2OCH_3}{\big<}}$

M − 75   Possible Precursor Compounds

Loss of $-$CH$_2$OCH$_2$CH$_2$OH

75   Structural Significance

Dimethyl acetals [(CH$_3$O)$_2$CH$-$]
Disubstituted benzene derivatives containing electrophilic
   substituents C$_6$H$_3$
Dichlorobutenes, CH$_2$ClCH=CH$-$, CH$_2$CH=CHCl
Propionates (generally small)
Sulfides (C$_2$H$_5$SCH$_2$$-$)
(CH$_3$)$_2$SiOH, $-$CF=CHCH$_2$OH, C$_2$H$_5$OC(O)+2H,
   (CH$_3$)$_2$CH$-$S$-$, CH$_3$OCH(OH)CH$_2$$-$

M − 76   Possible Precursor Compounds

Loss of C$_2$H$_4$SO from ethyl sulfones
Loss of CF$_2$CN

76   Structural Significance

Benzene derivatives (C$_6$H$_4$)
Aliphatic nitrates ($-$CH$_2$ONO$_2$)
Propyl thioethers (C$_3$H$_7$SH)
$-$CF$_2$CN, C$_6$H$_4$, CS$_2$, C$_3$H$_5$Cl, N(OCH$_3$)$_2$, (CH$_3$)$_2$NS†

M − 77   Possible Precursor Compounds

*Also observe *m/z* 74, 73, and 72.
†*m/z* 76, 42, and 61 suggest (CH$_3$)$_2$NS$-$.

77   Structural Significance

Monosubstituted benzene derivatives containing an electro-philic substituent.

$(CH_3)_2SiF$, $-CF_2CH=CH_2$
$C_3F_2H_3$ (e.g., $-CHFCF=CH_2$, $-CH_2CH=CF_2$, $CH_3CF=CF-$
$-CH=CHCHF_2$), C_6H_5-, $(CH_3)_2P(O)$, $CH_2ClC(O)$,
$C_2H_5Si(H)F$

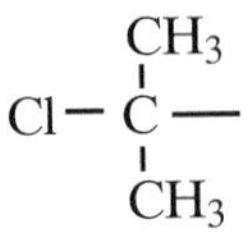

$$Cl-\overset{\displaystyle CH_3}{\underset{\displaystyle CH_3}{C}}-\qquad CH_3CH_2CHCl-$$

Alkylbenzene (m/z 39, 50, 51, 52, 63, 65, 76, 77, and 91)
$C_nF_{2n+1}CH_2CH_2I$

M − 78   Possible Precursor Compounds

Loss of benzene from compounds such as $(C_6H_5)_2$ $CHOCH(C_6H_5)_2$

78   Structural Significance

Phenyl tolyl ethers

Compounds containing the pyridyl group

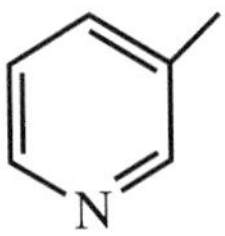

C_2F_2O, $C_3F_2H_4$,

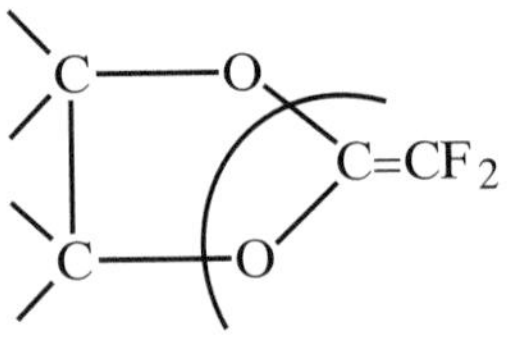

C_2H_7PO

M − 79   Possible Precursor Compounds

Bromides

79   Structural Significance

Cycloalkadienes
Bromides
Pyridines
$-CH=CClF$, CH_3SS-, $-CF_2C_2H_5$

$CH_3C_5H_4$, , $CH_3\overset{\overset{O}{\|}}{\underset{\underset{O}{\|}}{S}}-$, $CH_2ClCHOH$,

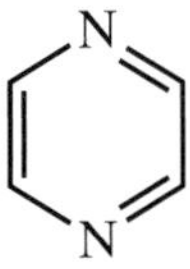

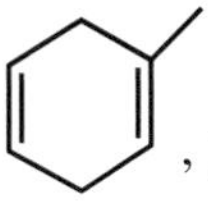

, Br

$CHF_2C(O)$, $HOP(O)OCH_3$

M − 80   Possible Precursor Compounds

Alkyl bromides (loss of HBr)

80   Structural Significance

Alkyl pyrroles
Substituted cyclohexenes (C_6H_8)
Methyl disulfides (CH_3SS+H)
C_5H_6N, C_2H_2ClF,

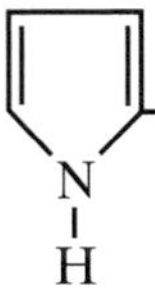

$-CH_2-$, $\longrightarrow$ C_5H_6N (*m/z* 80)

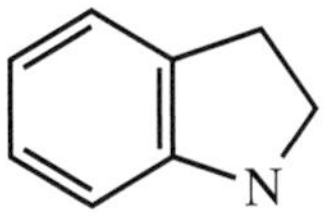

$NO_2-$$-NH_2$ $\longrightarrow$ C_5H_6N (*m/z* 80)

M − 81   Possible Precursor Compounds

81   Structural Significance

Hexadienes (C_6H_9)
Alkyl furans
N-TFA, *n*-butyl histidine

CH_3OCF_2-, CH_3SiF_2, $CF_2=CF$-, -CH=CH-CH=CH-CH=O,
-CF_2CH_2OH, CH_3CFCl-,

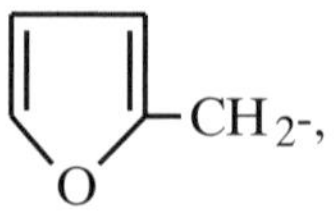

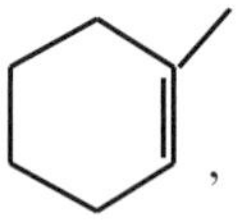

,

C_6H_9, $(HO)_2P+O$ present in phosphate spectra

M − 82   Possible Precursor Compounds

Loss of $C_3H_4N_3$,

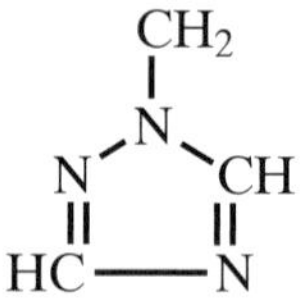

82   Structural Significance

Aliphatic nitriles ($-CH_2CH_2CH_2CH_2CN$) (*m/z* 82, 96, 110,
   124, 138, 152, etc., suggest straight-chain nitriles)
Benzoquinones
Piperidine alkaloids
Some fluoroalcohols
m/z 82 and 67 suggest cyclohexyl compounds (see *m/z* 83)
m/z 82 and 182 suggest cocaine (MW = 303)

CCl_2, C_6H_{10}, C_2F_3H, $CHF=CF_2$, $(CD_3)_3Si$-, C_5H_8N,

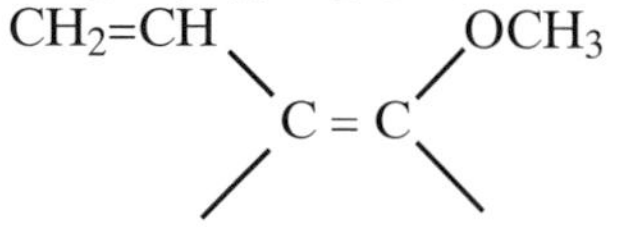

M − 83   Possible Precursor Compounds

Dicyclohexylbenzene

83   Structural Significance

Aliphatic nitriles
Cyclohexanes (C_6H_{11}) (*m/z* 83, 82, 55, and 41 suggest a cyclohexyl ring)

Thiophene derivatives
Trialkyl phosphites
SO_2F, $CHCl_2$, $C_2H_3F_3$,* CHF_2CHF-, $-CF_2CH_2F$,
CF_3CH_2-, C_5H_7O, C_5H_9N-, $(CH_3)_2C=CH-C(O)$ (e.g., mesityl oxide)†
$CH_3CH=CHCH_2C(O)$
$CH_3CH_2CH=CHC(O)$
$CH_2=CHCH_2CH_2C(O)$
$HP(OH)_3$, phosphonic acid derivatives

M − 84   Possible Precursor Compounds

Loss of $-N$ ⟨ring⟩ ($C_5H_{10}N$) observe *m/z* 84 and 56

84   Structural Significance

Aliphatic isocynate $-CH_2CH_2CH_2NCO$
Esters of aliphatic dibasic carboxylic acids (>dimethyl suberate)

Piperidines (*m/z* 84, 56, etc.) , $C_5H_{10}N$,

Pyrrolidines ,CH_2Cl_2, $C_5H_{10}N$, , $CDCl_2$,

C_3H_7-C=N-CH_3, C_4H_8Si, ,$C_4H_4O_2$

---

*m/z 83 and 33 suggest CF_3CH_2-.
†m/z 55 and 83 suggest this structure.

M − 85   Possible Precursor Compounds

Glycol diacrylates (M − CH$_2$=CHC(O)OCH$_2$−)
Loss of (C$_4$H$_9$ + CO) from TBDMS derivatives of amino acids
Caffeine [−C(O)N(CH$_3$)C(O)−]

85   Structural Significance

Butyl ketones (C$_4$H$_9$C(O))
N-TFA, *n*-butyl glutamic acid
Piperazines (C$_4$H$_9$N$_2$)
Tetrahydropyranyl ethers
Lactones
Methyl-8 derivatives of primary amines
SiF$_3$, CF$_2$Cl, CF$_3$O, POF$_2$, C$_4$H$_5$O$_2$, C$_5$H$_9$O, C$_5$H$_{11}$N, C$_6$H$_{13}$,
   (CH$_3$)$_2$NCH=NCH$_2$−

CH$_3$C(O)C(CH$_3$)$_2$ -,   —CH$_2$ — ,

-CH=CH-C(O)OCH$_3$

M − 86   Possible Precursor Compounds

Loss of CH$_2$=C(CH$_3$)C(O)OH from methacrylate esters

86   Structural Significance

Propyl ketones (C$_3$H$_7$C(O)CH$_2$ + H)
a-Methylisothiocyanates
SOF$_2$, C$_4$F$_2$, (C$_2$H$_5$)$_2$NCH$_2$−, C$_4$H$_9$NHCH$_2$, C$_5$H$_{12}$N
C$_4$H$_9$CHNH$_2$,   HC(O)NHC(CH$_3$)$_2$,   −CH(CH$_3$)N=C=S,
   −CH$_2$NHC$_4$H$_9$
C$_2$H$_5$NHCH$_2$CH$_2$CH$_2$− (will observe *m/z* 44, 58, 72, and 86)

M − 87   Possible Precursor Compounds

n-Amyl esters
Loss of −CH(CH₃)C(O)OCH₃
Loss of C(O)OC₃H₇ from *N*-TFA, isopropyl esters of amino acids

87   Structural Significance

Methyl esters of aliphatic acids (*m/z* 74 and 87 indicate a methylester)
Glycol diacetates (*m/z* 87 and 43 are characteristic ions)
Methyl dioxolanes
Long-chain methyl esters
Esters of *n*-chain dibasic carboxylic acids
CH₂CH₂C(O)OCH₃
C₄H₇O₂:

$$CH_3 \diagdown$$
$$CHOCH \diagup$$
$$CH_3 \diagup \quad \diagdown CH_3$$

$$CH_3C(O)OCH \diagup CH_3$$
$$\diagdown$$

CH₃CHCH₂CO₂H
  |

CH₃C(O)OCH₂CH₂, C₂H₅C(O)OCH₂−C₅H₁₁O:
  CH₃CH₂CH(CH₃)OCH₂−, −CH(OH)CH₂CH(CH₃)₂,
  −CH₂OCH₂CH(CH₃)₂

C₄H₄Cl, C₄H₇S, [(C₂H₅)₂NCH₂+H], C(O)OC₃H₇,

M − 88   Possible Precursor Compounds

88   Structural Significance

Long-chain ethyl esters (look for *m/z* 101)
(C₂H₅OC(O)CH₂+H)

$C_3H_6NO_2$, CF_4, $C_4F_2H_2$, $-C(CH_3)_2NCO$
$H+CH(CH_3)C(O)OCH_3$, $C_4H_{10}NO$, C_4H_8S, $-CHNH_2-$
$C(O)OCH_3$
Long-chain methyl esters with a 2-methyl group

M − 89   Possible Precursor Compounds

Loss of $OSi(CH_3)_3$ from TMS derivatives (m/z 90 should be more abundant)

89   Structural Significance

Triethylene glycol (also look for m/z 45)
O- and N-containing heterocyclic compounds
Characteristic rearrangement ion of butyrates except methyl
Dinitrotoluenes
Trimethylsilyl derivatives
Sulfur-containing compounds (also expect m/z 61)

C_7H_5, C_4H_6Cl, $C_3H_7SCH_2-$, , $C_4F_2H_3$,

$(CH_3O)_2C\overset{\diagup CH_3}{\diagdown}$

C_4H_9S-, $C_3H_7SCH_2-$, $CH_2=CHCF=CF-$,
$-CH_2OSiH(CH_3)_2$
$-CH(OH)CH_2CH(OH)CH_3$, $(CH_3)_3SiO-$, $[C_3H_7OC(O)+$
$2H]$, $(CH_3O)_2CCH_3$, $CH_3OSi(CH_3)_2$
$-CF=CHCH_2OCH_3$ (should observe m/z 45 and 59 as well)

M − 90   Possible Precursor Compounds

Loss of $(CH_3)_3SiOH$ from TMS derivatives (may see the loss of 180 and/or 270 Daltons from the molecular ion, depending on the number of OH groups present).

90   Structural Significance

O- and N-containing heterocyclics
Aliphatic nitrates with an α-methyl group (CH_3CHONO_2)
Methyl esters of 2-hydroxy carboxylic acids

C_6H_4N, C_3FC, C_6H_5CH, C_7H_6, $(C_2H_5)_2PH$,

M − 91   Possible Precursor Compounds

91   Structural Significance

Alkyl benzenes (C_7H_7) (*m/z* 104 and/or 117 are also characteristic ions)
Phenols
Aromatic alcohols
Benzyl esters
Alkyl chlorides $(C_6\text{-}C_{18})$; C_4H_8Cl from terminal chloroalkanes
N-TFA, *n*-butyl phenylalanine

$CH_2ClCH_2C(O)$,

$ClCH_2CHCH_3CH_2-$, $C_4H_5F_2$, $Cl(CH_3)_2CCH_2-$
$CF_2=C(CH_3)CH_2-$, $-CF_2C(CH_3)=CH_2$, *m/z* 43 and 91
suggest 1-chlorodecane

$$CH_3CHCH=CF_2, C_2H_4O_2P$$
$$|$$

M − 92   Possible Precursor Compounds

Certain esters of dibasic carboxylic acids
Loss of (C_3H_9SiF) for fluorinated sugars

Loss of   —NH—

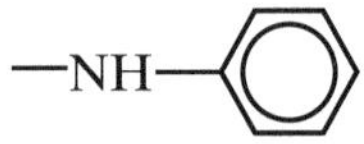

92   Structural Significance

β and γ-monoalkyl pyridines

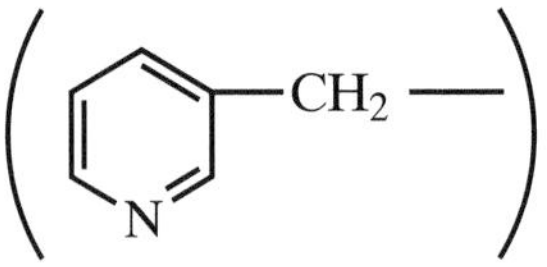

Phosphorus compounds
Salicylates (see *m/z* 138)
Benzyl compounds with a γ-hydrogen

(CH_3)$_3$SiF, N_3CF_2-

C_7H_8 double rearrangement ion common to alkylbenzenes
CFClCN
C_3H_8PO

M − 93   Possible Precursor Compounds

Phenoxy derivatives

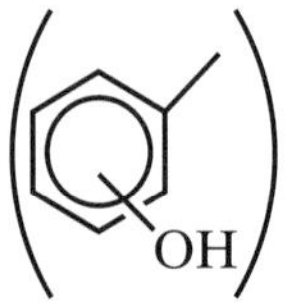

93 Structural Significance

Nitrophenols
Fluorocarbons
Salicylates (see *m/z* 138)

Alkyl pyridines 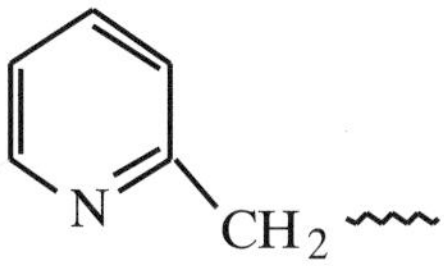 + H, 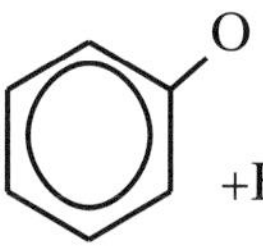 +H,

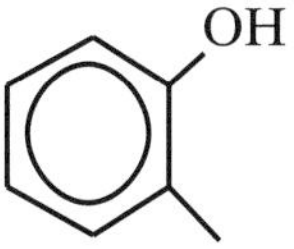 , C_3H_3FCl: $-CFClCH{=}CH_2$, $CH_3OCF{=}CF-$

C_3F_3: $CF_3-C{\equiv}C-$

$(CH_3O)_2P$, $-C(O)OCH_2Cl$,

C_3H_6OCl: $C_2H_5OCHCl-$, 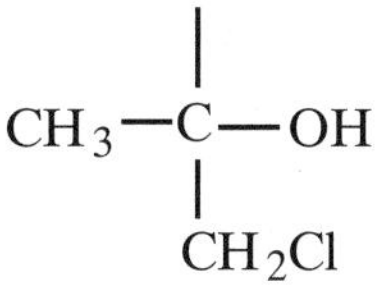

C_2H_6SiCl: 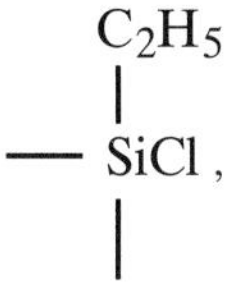 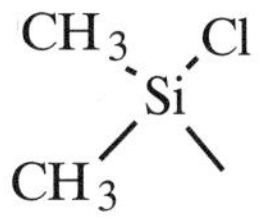

M − 94 Possible Precursor Compounds

94 Structural Significance

Alkyl phenyl ethers (except anisole)
Dimethyl pyrroles

C_6H_5OH from benzopyrans 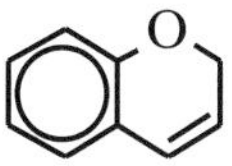

Alkylpyrazines

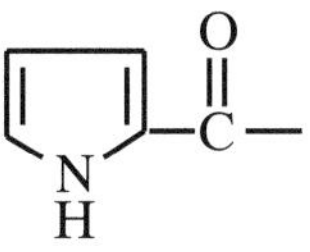 , C_2Cl_2, C_3F_3H, $CH_3OP(OH)CH_3$,

CH_3Br, $ClCC(O)F$ from

M − 95 Possible Precursor Compounds

Methyl esters of aromatic sulfonic acids (SO_2OCH_3)

95 Structural Significance

Derivatives of furan carboxylic acids

Hydroxymethylcyclohexanes
Methylfurans
Methyl esters of sulfonic acids
Segmented fluoroalcohols: $C_nF_{2n+2}CH_2CH_2OH$ (*m/z* 31, 95, 69, and 65)

$-S\text{-}OCH_3$, C_7H_{11}, $CF_3C=CH_2$, C_6H_9N, CF_3CN,

$-CF_2CH_2CH_2OH$, $CHF_2CF=CH-$, C_6H_4F, C_6H_7O, C_3H_5ClF,

$C_5H_7N_2$, CH_2-, $C_2F_2O_2H$,

$CF_3CH=CH-$, $C_3F_3H_2$, F——

M − 96 Possible Precursor Compounds

96 Structural Significance

Aliphatic nitriles $(-(CH_2)_5CN)$ (*m/z* 96, 82, and 110 suggest nitriles)

Piperidines
Esters of dibasic carboxylic acids
Dicycloalkanes
$(CH_3)_2SiF_2$, C_6H_5F, CH_3CH_2CFCl-

C_6H_8O:

M − 97   Possible Precursor Compounds

Loss of $(CF_2=CFO)$ from fluoroesters, fluoroamides, and
  fluoronitriles
(e.g., $CF_2 = CFOCF_2CF(CF_3)OCF_2CF_2C(O)OCH_3$
      $CF_2 = CFOCF_2CF(CF_3)OCF_2CF_2C(O)NH_2$
      $CF_2 = CFOCF_2CF(CF_3)OCF_2CF_2CN)$

97   Structural Significance

Alkyl thiophenes

O- and F- containing compounds: $[CF_3C(O), -CF_2C(O)F,$
$-CF-CF_2$, $CF_2=CF-O-]$

Aliphatic nitriles $(C_6H_{11}N)$

$CF_3-N=N-$, $CFCl=CF-$,

$CH_2=CHCH_2CH(CH_3)C-$, C_7H_{13}, C_2F_2Cl

CH_3CCl_2-, $CH_3CH_2CH_2CH=CHC(O)$
$CH_3P(O)Cl$, C_7H_{13}, $CH_3P(OH)_3$, $CH_2ClCHCl-$

M − 98   Possible Precursor Compounds

Loss of -(C$_7$H$_{14}$)

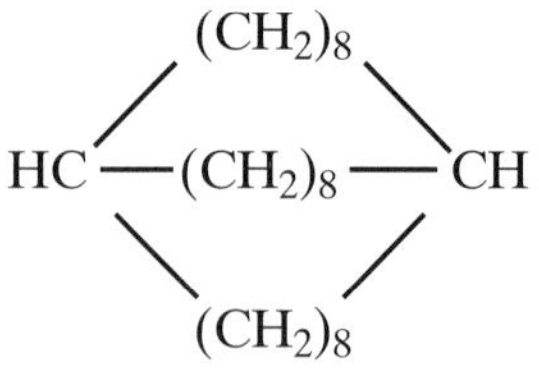

$$HC \overset{\displaystyle (CH_2)_8}{\underset{\displaystyle (CH_2)_8}{-\!\!-(CH_2)_8-\!\!-}} CH$$

98   Structural Significance

Piperidine alkaloids (*N*-alkyl)

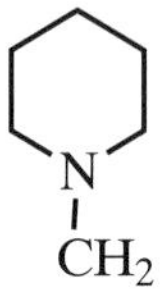

Dicarboxylic esters
Alkyl thiophenes
Dicarboxylic acids (e.g., palmitic and fumaric acids)
Bis(hexamethylene)triamine

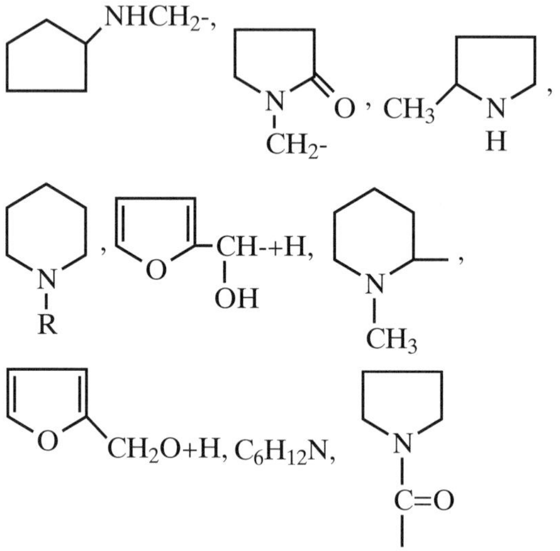

CH_2O+H, $C_6H_{12}N$,

M − 99   Possible Precursor Compounds

Loss of (CD$_3$)$_3$SiOH from deuterated TMS derivatives

99   Structural Significance

Maleates (>methyl)*

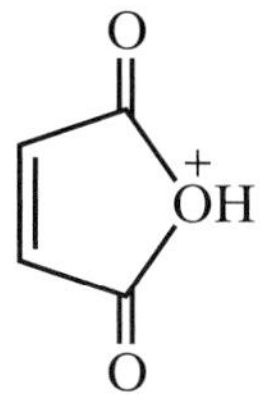

N-TFA, *n*-butyl aspartic acid
Acrylates (*m/z* 99 and 55 suggest glycol acrylates)
Isocyanates (C_5H_9NO) (also look for *m/z* 56)
Ethylene ketals of cyclic compounds (e.g., steroids)
Amyl ketones ($CH_3(CH_2)_4C(O)$)
Pentafluorobenzene
σ-Lactones, $CH_2{=}CH{-}C(O)OCH_2CH_2$
$CF_3OCH_2{-}$

CF_3CHOH, C_5F_2H, $CH_3C(O)C_2H_4C(O)$, C_5H_9NO,

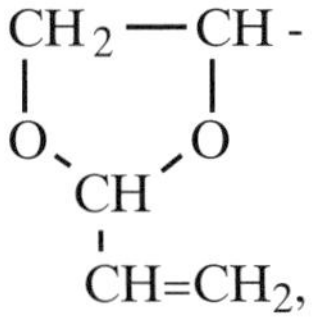

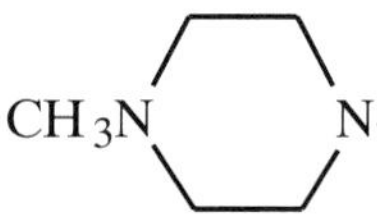

$C_6H_{13}N$,

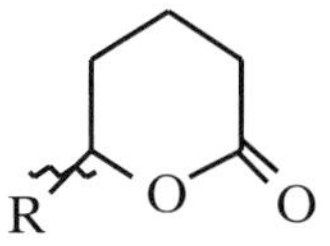

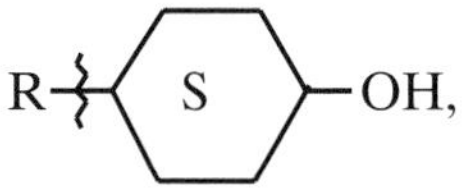

$CH_2{=}CHC(O)NHCH_2NH{-}$,

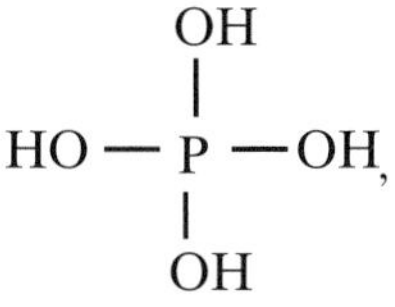

(Methyl-8 derivatives of amides) → $-C(O)N{=}CH{-}N(CH_3)$

M − 100   Possible Precursor Compounds

Loss of C_2F_4 from fluorine compounds

---

*Look for a fragment ion at *m/z* 29 and an ion at *m/z* 99 + 28 = 127 for diethyl-, a fragment ion at *m/z* 41 and *m/z* 99 + 40 = 139 for dipropyl-, and so forth.

100   Structural Significance

Perfluoroalkenes and perfluorocycloalkanes (C_2F_4)

$(C_2H_5)_2NC(O)$, $-C_5H_{11}CHNH_2$, $C_4H_9C(O)CH_2+H$,

+ H,

C_5H_5Cl, $(CH_3)_2SiN_3$, $(CH_3)_2CHN(CH_3)C(O)$
m/z 86, 100, 114, 128, 142, 156, etc. suggest diamines. Look for a large *m/z* 30 peak.

M − 101   Possible Precursor Compounds

Loss of $-C(CH_3)_2C(O)OCH_3$
N-TFA, *n*-butyl amino acids
Loss of $(-C(O)OC_4H_9)$

101   Structural Significance

Ethyl esters $(-CH_2CH_2C(O)OC_2H_5)$
Butyl esters $(-C(O)OC_4H_9)$
Succinates (e.g., dicyclohexyl succinate)
Hexafluoropropylene oxide (HFPO) dimers and trimers
Fluorochloro compounds having $CFCl_2$
Malonic acids $C(O)CH_2C(O)OCH_3$
Cyclic sulfides

$C_4H_5O_3$, CF_3CFH-, $CFCl_2$, CHF_2CF_2-, $C_6H_{13}O$, $C_5H_9O_2$,
$CH_3(CH_2)_4CHOH-$, $CH_3SC(O)CH=CH-$,
$-CH_2CH_2OC(O)C_2H_5$, $CH_3C(O)OCHC_2H_5$,

$$C_3H_7 - \overset{\overset{\displaystyle CH_3}{|}}{\underset{|}{C}} - CH_2OH, \ PSF_2, \ CF_3S, \ PCl_2,$$

$$CH_3OC(O) - \overset{\overset{\displaystyle CH_3}{|}}{\underset{\underset{\displaystyle CH_3}{|}}{C}} - \ ,$$

M − 102   Possible Precursor Compounds

Loss of acetic anhydride from sugar acetates

102   Structural Significance

Quinolines
Long-chain propyl esters ($C_3H_7OC(O)CH_2+H$)

NC—⟨benzene ring⟩ , $C_6H_5C≡CH$ (e.g., phenylmaleic anhydride)

$CH_3OC(O)C(CH_3)_2+H$
  (e.g., $CH_3OC(O)C(CH_3)_2C(O)OCH_3$)
$(CH_3)_3SiNHCH_2−$, $−CHNH_2−C(O)OC_2H_5$

M − 103   Possible Precursor Compounds

Loss of $[CH_2OSi(CH_3)_3]$ from molecular ion of TMS

Ketohexoses and 1,2,6-hexanetriol

103   Structural Significance

Alkyl indoles (characteristic rearrangement ion)
Cinnamates ($C_6H_5CH=CH−$)
Valerates (>methyl)
Double rearrangement of protonated carboxylic acids
Trimethylsilyl derivatives $[CH_2OSi(CH_3)_3]$
1,1-Diethoxyalkanes $[(C_2H_5O)_2CH−]$

$(CH_3)_2CHSCH_2CH_2-$, C_8H_7, $C_4H_9OC(O) +2H$,

$CH_3OC(O)CH_2\overset{\bullet}{C}HOH,$* $CH_3CHCH_2OCH_2CH_2-$,
                         |
                         OH

$CH_3CH_2C(OCH_3)_2-$

$$CH_2-\overset{\diagup}{C}-Cl$$

*From dimethyl malate.

M − 104   Possible Precursor Compounds

104   Structural Significance

Tetralins (tetrahydronaphthalene)
m/z 104 and 158 (see phenylcyclohexene)

CF_3Cl, N_2F_4, SiF_4, C_8H_8, $C_2H_5CHONO_2$,

—CH_2—⟨◯⟩—CH_2—, $CH_3SCH_2CH_2CHNH_2$-,

or

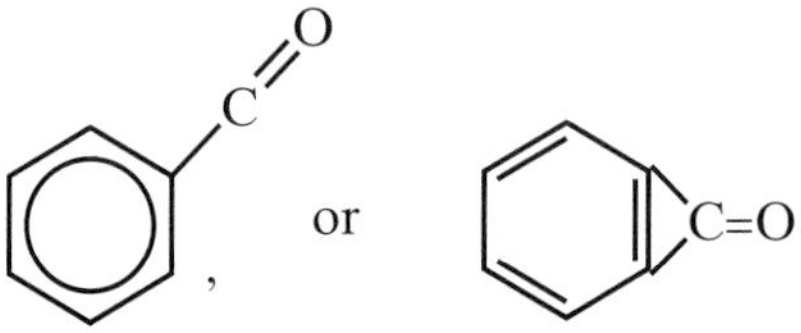

m/z 104 and 91 or 117 are characteristic of some alkybenzenes.

M − 105   Possible Precursor Compounds

Esters of aliphatic dibasic carboxylic acids
Benzoin ethers (e.g., benzoin isopropyl, benzoin isobutyl)

Loss of CH=C —⟨◯⟩—CH_3 from

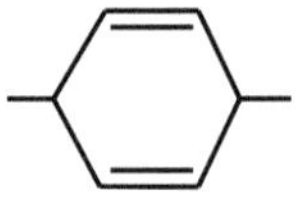

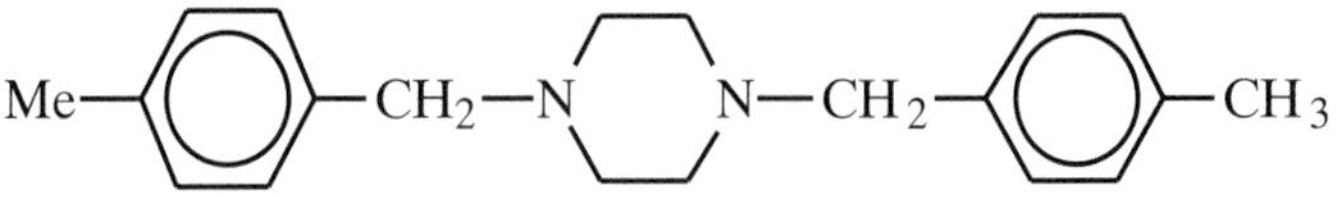

Me—⟨◯⟩—CH_2—N⟨ ⟩N—CH_2—⟨◯⟩—CH_3

Loss of $[CH_3+(CH_3)_3SiOH]$ from TMS derivatives of aldohexoses, etc.

105   Structural Significance

Benzoates
Aromatic alcohols
Azobenzenes
$(C_2H_5)_2SiF$, $C_6H_5C=O$, $C_6H_5N=N$, C_4F_3, $(CH_3O)_2SiCH_3$

$C_6H_5CHCH_3$, $C_6H_5CH_2CH_2$, C_7H_7N,

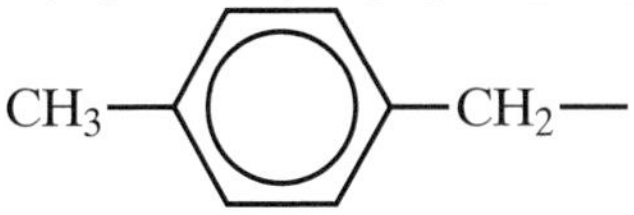

CH_3-⟨○⟩$-CH_2-$, $CHCl=\overset{|}{C}\text{-}COOH$

M − 106   Possible Precursor Compounds

Esters of long-chain dibasic carboxylic acids

106   Structural Significance

Substituted *N*-alkylanilines ($C_6H_5NHCH_2-$)

Alkyl pyridines: ⟨N⟩$-CH_2CH_2-$,

CH_3⟨pyridine⟩CH_2- ,

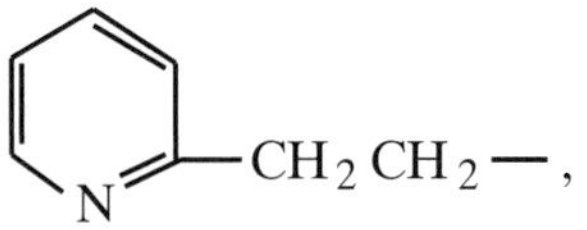

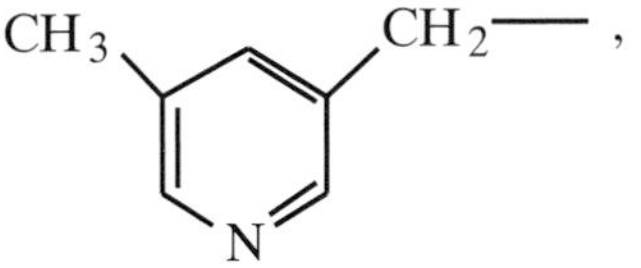

C_7H_8N, C_4F_3H,

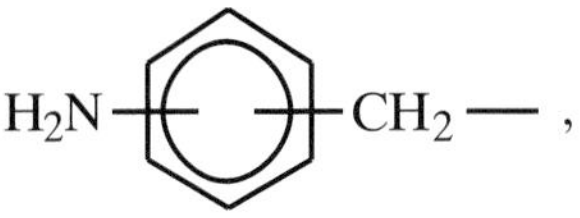

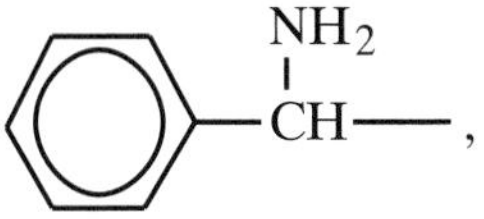

H_2N-⟨○⟩$-CH_2-$,   ⟨○⟩$-\overset{NH_2}{\underset{}{CH}}-$,

$(CH_3)_2NP(O)CH_3$,   ⟨○⟩$-NHCH_2-$,

⟨N⟩$-\overset{O}{C}\text{=}$

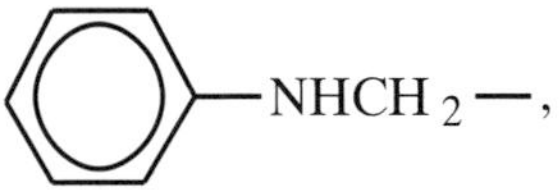

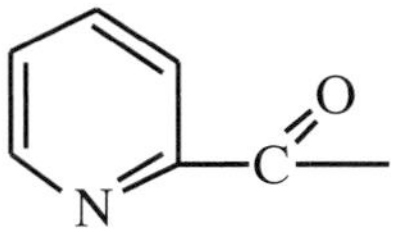

M − 107   Possible Precursor Compounds

Loss of ($CH_3C_6H_4O-$) as in ($CH_3C_6H_4O)_3P$
($CH_3+C_3H_9SiF$) from TMS derivative of fluorinated sugars

e.g., CF_3—$CHOH$

CF_3—C—OH

CF_3—C—OH

CF_3 $CHOH$

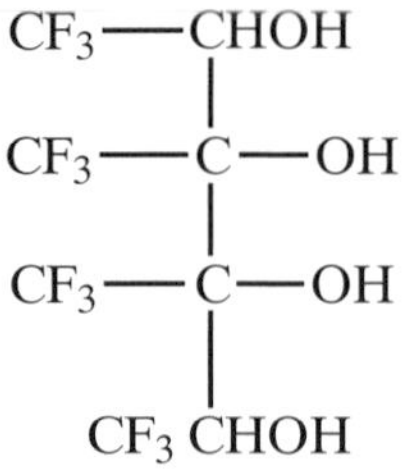

107 Structural Significance

Alkyl phenols ($HOC_6H_4CH_2$—), pyrethrin II (MW = 372)
See benzoin isopropyl ether (MW = 254)

$C_6H_4CH_2O$-, [benzene ring] $CHOH$-, $-CF_2C_4H_9$,

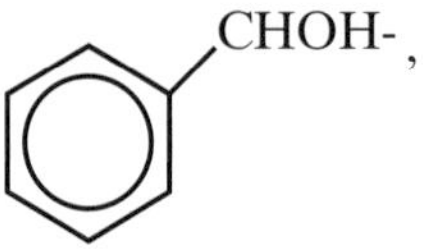

$-CH_2CH_2OCH_2CH_2Cl$, $C_6H_5SiH_2$

$CH_3(CH_2)_3CH=CH-C\equiv C-$
  or $CH_3(CH_2)_3-C\equiv C-CH=CH-$
$C_6H_5OCH_2$—, $CH_3C_6H_4O$—, $-C_6H_4OCH_3$

M − 108 Possible Precursor Compounds

Loss of
$$\text{H}-\overset{\overset{\displaystyle Ph}{|}}{\underset{\underset{\displaystyle H}{|}}{Si}}-\text{H} \quad \text{from} \quad \text{H}-\overset{\overset{\displaystyle Ph}{|}}{\underset{\underset{\displaystyle H}{|}}{Si}}-\overset{\overset{\displaystyle Ph}{|}}{\underset{\underset{\displaystyle H}{|}}{Si}}-\overset{\overset{\displaystyle Ph}{|}}{\underset{\underset{\displaystyle H}{|}}{Si}}-\text{H}$$

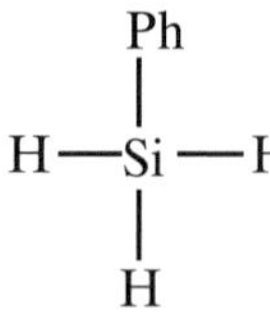

—NH—[benzene ring]—OH

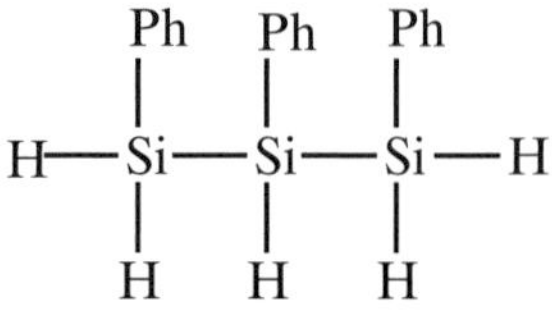

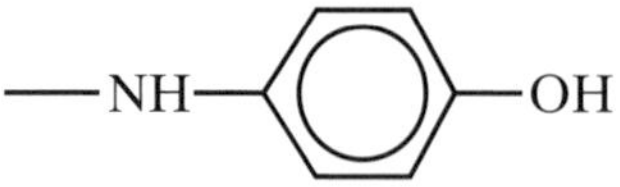

108 Structural Significance

Tolyl ethers ($CH_3C_6H_4O$+H)
Alkyl pyrroles
Benzothiazoles (C_6H_4S)

$H_2NC_6H_4O\text{-}, C_6H_5SH, C_6H_5CH_2O\text{-+}H, C_6H_5P{\diagup}^{*}{\diagdown}$

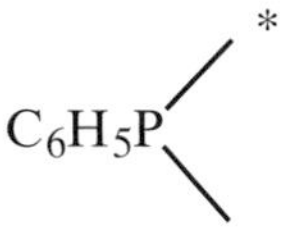

$C_6H_5CH_2OH,$

OH

CH$_2$-

, $-\overset{\displaystyle O}{\underset{\displaystyle HN-CH_3}{P}}-OCH_3$

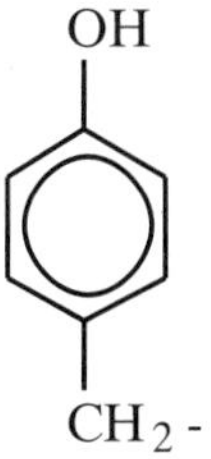

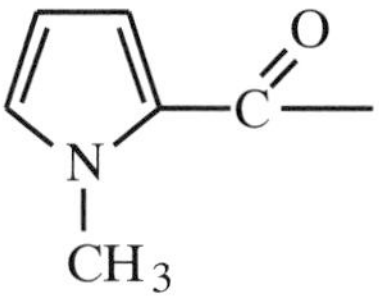

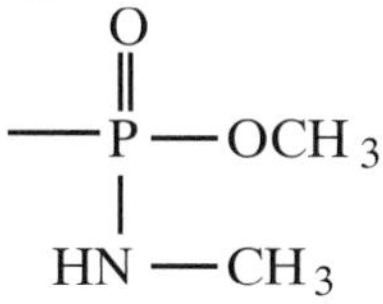

M − 109   Possible Precursor Compounds

109   Structural Significance

Purines ($C_5H_7N_3$)
Certain perfluoroketones (C_3F_3O)

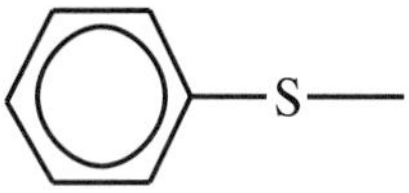

—S—— , C_3F_2Cl, $HOC_6H_4O\text{-}$, $CH_3OCF=CFO\text{-}$,

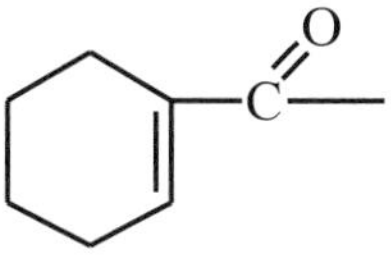

, $CF_2=CFC(O)$, $C_4H_4F_3$, C_8H_{13}, C_3F_2Cl,

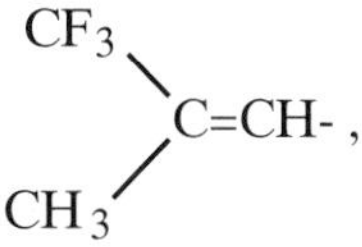

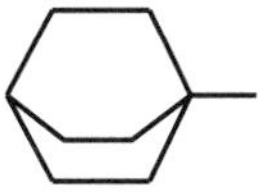

$\underset{\displaystyle CH_3}{\overset{\displaystyle CF_3}{}}C=CH\text{-}$, , $CH_3CCl=CCl\text{-}$,

—CH$_2$—— , HO——NH - +H,

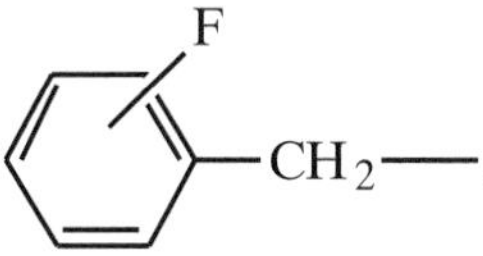

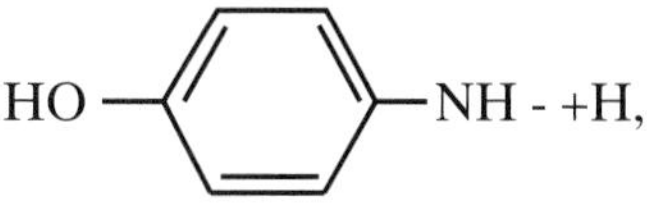

*See triphenylphosphine, MW = 262.

$$-CCl_2CH=CH_2,$$

$$HOP(O)OC_2H_5$$

M − 110   Possible Precursor Compounds

110   Structural Significance

Esters of aliphatic dibasic carboxylic acids
Aliphatic cyanides* $(CH_2)_6CN$, $C_7H_{12}N$,
C_6F_2, C_2Cl_2O, $C_7H_{10}O$, $CH_3OPOH(O)CH_3$, $C_4H_5F_3$,
   $CH_2=C(CF_3)-CH_3$,

M − 111   Possible Precursor Compounds

Suggests an octyl group (C_8H_{16})

111   Structural Significance

Methyl alkyl thiophenes
Adipates, $C_6H_7O_2$
Esters of aliphatic dibasic carboxylic acids
Long-chain methyl esters of carboxylic acids

$$CH_3CH_2CCl_2-, \quad -C_6H_5Cl,$$

$$C_8H_{15}, \quad C_7H_{11}O,$$

$CFBrH$, $CF_3CH(CH_3)CH_2-$, $C_3H_5Cl_2$
$C_2H_5OP(O)C_2H_5$, $CHCl_2C(O)$, $-CF_2CH_2C(O)F$
$CF_3C(O)CH_2-$, $CF_3C = N-NH_2$,

$$\overset{\text{N-N(CH}_3)_2}{\underset{\parallel}{}}$$
$$-CH=CH-\overset{}{C}-CH_3$$

$$-CH_2CH_2-,$$

*m/z 110, 96, and 82 suggest nitriles.

CH$_3$ | (cyclohexyl)—CH—, -CF$_2$C(Cl)=CH$_2$, C$_2$H$_5$P(OH)$_3$

M − 112   Possible Precursor Compounds

Loss of (piperidine N-C=O)

R′−H (suggests an octyl group)
Loss of (CH$_2$)$_5$NCO from isocyanates

112   Structural Significance

N-alkylcyclohexylamines
Dibutylhexamethylenediamine
2,7-Dioxo-1,8-diazacyclotetradecane
C$_7$H$_{14}$N, C$_3$F$_4$, C$_6$H$_{12}$Si, C$_6$H$_5$Cl, C$_6$H$_8$O$_2$, (CH$_3$)$_2$P(O)Cl

M − 113   Possible Precursor Compounds

113   Structural Significance

Methyl esters of Δ^2-fatty acids
−CH$_2$CH$_2$CH=CHC(O)OCH$_3$
Methacrylates (CH$_2$=C(CH$_3$)−C(O)OCH$_2$CH$_2$−)

Certain diketones
N-methylsuccinimide
Cyclic ethers ($C_7H_{13}O$)
$C_4H_9C(O)CH_2CH_2$-, $(CH_3)_2CHC(O)CH_2C$-,
$CH_3CH_2C(O)CH_2CH_2C(O)$

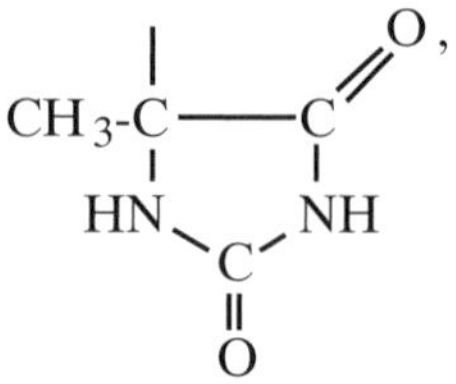

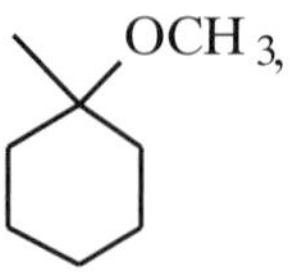

$CH_3C(O)CHCH_2C(O)CH_3$, -$CH_2C(CH_3)=CH$-$C(O)OCH_3$,
$CF_3CH=CF$-, -$CF_2CH=CF_2$, C_2FCl_2, $CH_2ClCHClCH_2$-,
C_2F_2ClO, C_3F_4H, $C_6H_3F_2$, $C_6H_{13}C(O)$, C_8H_{17},
$C_7H_{15}N$, $C_6H_{13}N_2$, $C_6H_{11}NO$, $C_6H_9O_2$, $CF_3C(O)NH_2$,

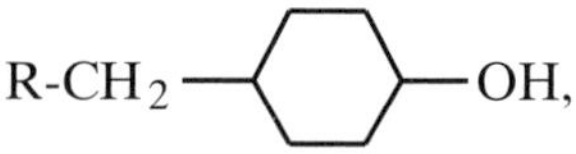

R-CH_2 ⬡ -OH, $C_7H_{13}O$, CH_3SiCl_2

$CF_2=CFS$-, $CH_2=CCH_3C(O)NHCH_2NH$-

M − 114	Possible Precursor Compounds
114	Structural Significance

Steroid alkaloids:

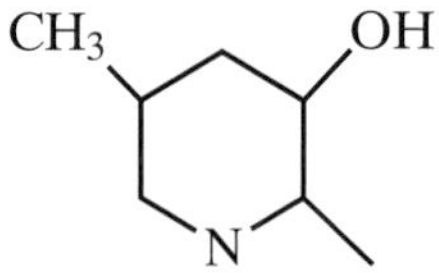

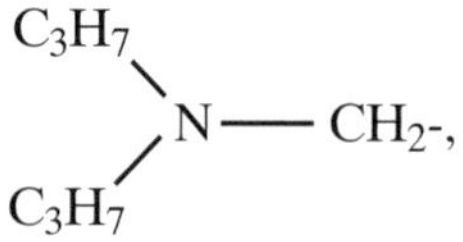

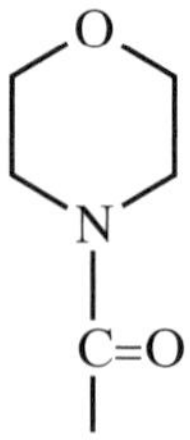

$C_7H_{16}N$, $C_6H_4F_2$, $CHCl$-$CClF$, $CF_3C(O)$-OH, -$CF_2N=CF_2$,
-$CH_2CH_2C(=NOH)C_3H_7$, -CF_2-$CF=NF$, C_2F_4N

M − 115	Possible Precursor Compounds

115   Structural Significance

Sugar acetates
Glutarates
Alkyl indoles (other than methyl)
C_9H_7 ion from naphthalenes (may produce a double-charged
    ion at m/z 57.5)
Creatinine-TRITMS (MW = 329)
Quinolines (C_9H_{11})

$-C(O)CH_2-C(O)OC_2H_5$, $C_2H_2FCl_2$, $CHCl_2CHF-$, $C_7H_{15}O$,
$CFCl_2CH_2-$, $(C_2H_5)_3Si-$, $(CH_3)_2CH-C(O)OCH_2CH_2-$,
$(CH_3)_3CSi(CH_3)_2-$, $CH_3CHCH=C(OH)OC_2H_5CH_2C(O)-OCH_3$,
$\overset{|}{CH_2-C(O)-}$

M − 116   Possible Precursor Compounds

Loss of $CH_2=CHC(O)O(CH_2)_2OH$ from some acrylates

116   Structural Significance

Amino alcohols and ethers ($C_6H_{14}NO$)
N-TFA, *n*-butyl glutamic acid

$-CH_2C_6H_4CN$, C_2F_3Cl, $-CF_2CFCl$, ⟨phenyl-S⟩— SH,

$C_3H_7C(O)-OC_2H_5$, C_5H_5OCl, $\sim\!\!\sim\!\!\sim\!\!\sim\!\!\sim$ $\begin{matrix} CF_2-CFCl \\ | \quad\quad | \\ CF_2-CFCl \end{matrix}$,

$C_6H_5\overset{|}{CHCN}$

M − 117   Possible Precursor Compounds

Loss of CCl_3 (e.g., DDT)
Loss of $-COOTMS$
Loss of CF_3CHFO from $CF_3CHFO|CF_2CF(CF_3)OCF_2CF_2CN$

117   Structural Significance

Alkyl indans
Styrenes (C_9H_9), caproates (>methyl)
CCl_3, C_2F_4OH, C_5F_3, $C_6H_5CH_2CH=CH-$,
   $CH_3C_6H_4CH=CH-$, $-CF_2OCHF_2$, $-CF_2CHClF$, $POCl_2$,
   $C_6H_5CH=CH-CH_2-$, C_2HF_3Cl, $CF_2ClCHF-$,
   $CH_3OCHCH_2C(O)OCH_3$, C_2HF_4O, $-C(O)OSi(CH_3)_3$,
   $-CH_2CH_2OSi(CH_3)_3$
Intense *m/z* 117 and 104 (characteristic of alkyl benzenes)
If one sees *m/z* 117 as a TMS derivative, it probably indicates
   that a carboxyl group is present.

M − 118   Possible Precursor Compounds

118   Structural Significance

$C_4H_9OCH_2CH_2OH$, $(CH_3)_2NSi(CH_3)_2O-$

M − 119   Possible Precursor Compounds

119   Structural Significance

Fluorocarbons (CF_3CF_2-)
Toluates ($CH_3C_6H_4C=O$)

Alkyl benzenes (e.g.,)

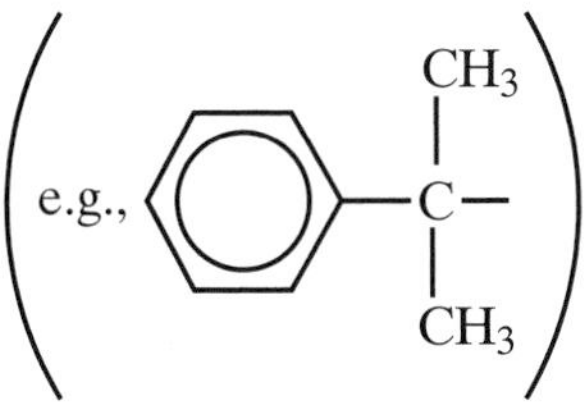

Pyrrolizidine alkaloids ()

— CH$_2$—⟨benzene⟩—CHO, CH$_3$—⟨benzene⟩—CHO

CH$_3$C(O)—⟨benzene⟩—

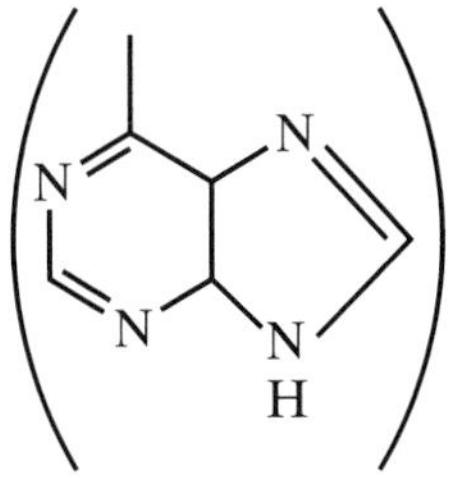

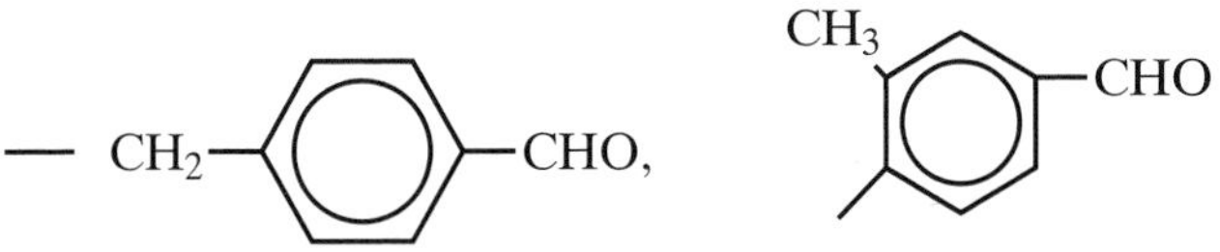

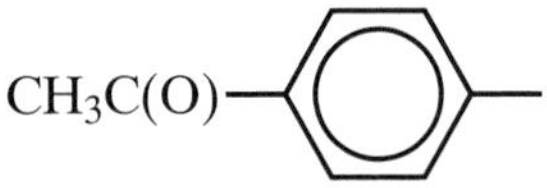

M − 120   Possible Precursor Compounds

120   Structural Significance

Salicylates
Pyrrolizidine alkaloids
Flavones
Isoflavones

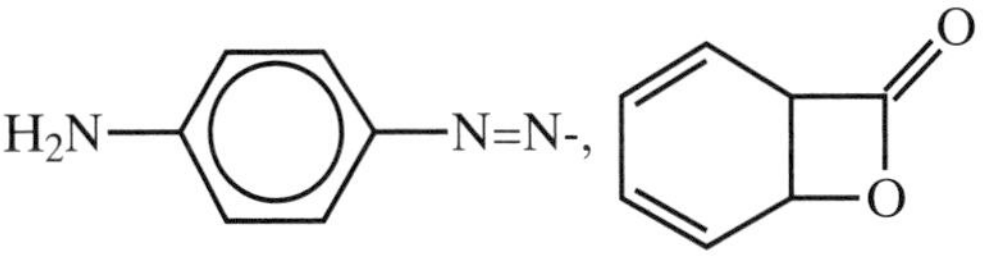

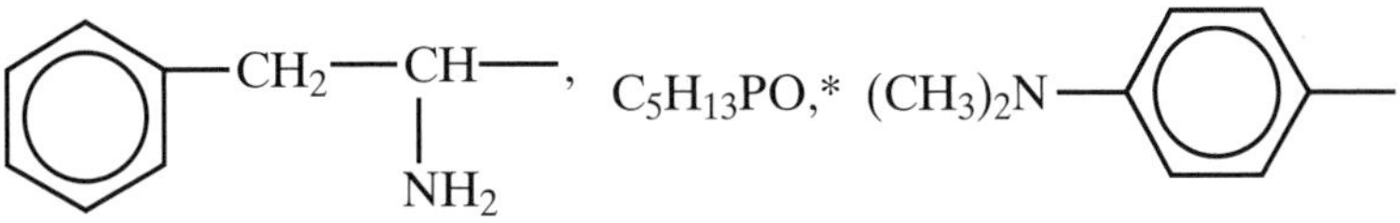

H$_2$N—⟨benzene⟩—N=N-,

⟨benzene⟩—CH$_2$—CH—, C$_5$H$_{13}$PO,* (CH$_3$)$_2$N—⟨benzene⟩—
                     |
                     NH$_2$

*From tributylphosphine oxide.

$$H_2N-\underset{C_6H_5CH_2CHNH_2}{\bigcirc}-C\equiv O$$

M − 121   Possible Precursor Compounds

121   Structural Significance

Salicylates (also see 138)

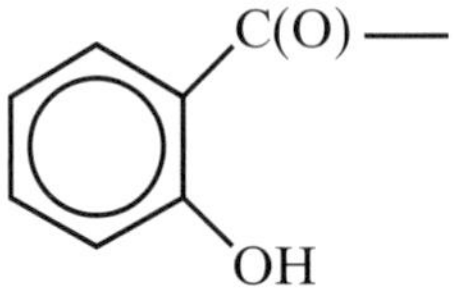

Alkyl phenols
Terpenes
$CH_3OC_6H_4CH_2-$, C_8H_9O, $(CH_3O)_3Si-$, $(C_2H_5)_2SiCl$,
$(CH_3)_2CBr$

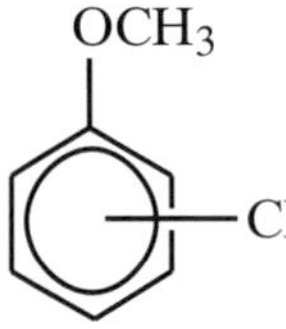

$$\underset{}{\overset{OCH_3}{\bigcirc}}-CH_2\text{-}, C_3H_7S\text{-}CHSH, C_6H_5CHOCH_3$$

$(CH_3O)_2Si(CH_3)O\text{-}$, $\underset{(CH_3)_2N}{\overset{CH_3NH}{\diagdown}}P\overset{O}{\diagup}-$

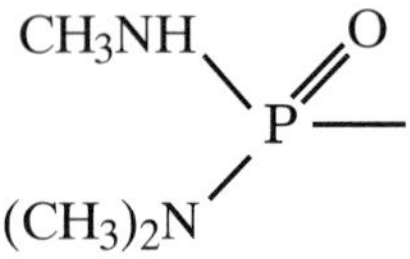

$HOC_6H_4CHCH_3$

M − 122   Possible Precursor Compounds

122   Structural Significance

$C_6H_5CO_2 + H$, Δ^3-ketosteroids

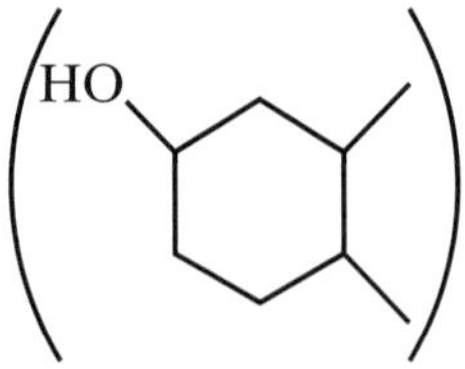

M − 123   Possible Precursor Compounds

123   Structural Significance

Pyrethrin I
Benzoates (by double hydrogen rearrangement) (The ester must be ethyl or higher.)
Trichlorobutenes

$CF_3CH=CHC(O)-$, $(HO)_2C_6H_3CH_2-$, $C_6H_5CH_2S-$

$CH_3C_6H_4S-$, $FC_6H_4C(O)-$, $C_8H_{11}O$, $C_2H_3SO_4$,

$-(CH_2)_2SO_3CH_3$,   , $C_6H_5SCH_2-$,

$CH_2FC_6H_4CH_2$, $-OS(O)_2OCH_2CH-$

M − 124   Possible Precursor Compounds

124   Structural Significance

Suggests an alkaloid in certain cases
Aliphatic cyanides (also may see *m/z* 96, 110, 138, 152, etc.)
Testosterone (also may see *m/z* 124, 288, and 246)

C_4F_4, $(CH_3O)_2P(O)CH_3$, $C_8H_{12}O$,

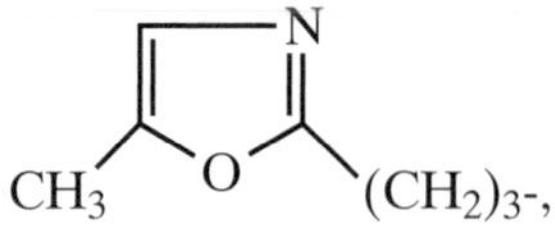

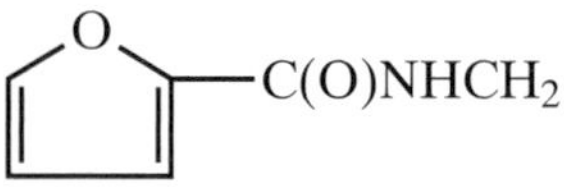

M − 125   Possible Precursor Compounds

125   Structural Significance

C_6H_5SO (e.g., the base peak in diphenylsulfone) (should also observe *m/z* 141)
C_4F_4H, $ClC_6H_4CH_2-$, C_6H_5CHCl-, $(CH_3O)_2P{=}S$, $C_7H_9O_2$,

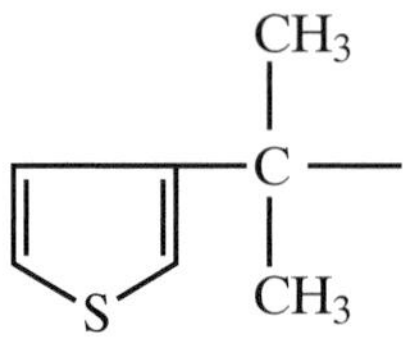

M − 126   Possible Precursor Compounds

126   Structural Significance

$CF_3C(O)NHCH_2-$,

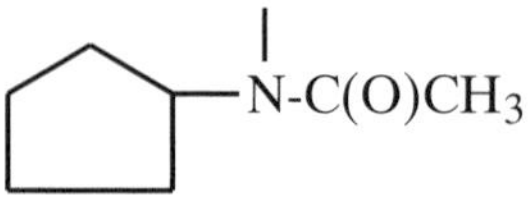

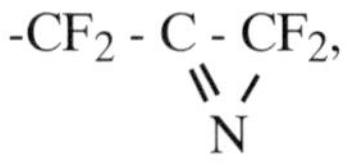

CF_3CFCN, $-CF_2 - C - CF_2$, $-CF_2CF_2CN$, $C_4F_4H_2$,

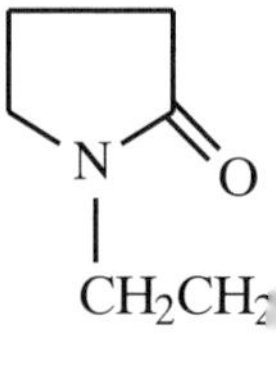

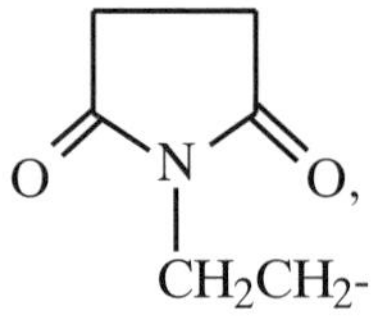

M − 127   Possible Precursor Compounds

Iodides

127 Structural Significance

Iodo compounds

Naphthyl compounds 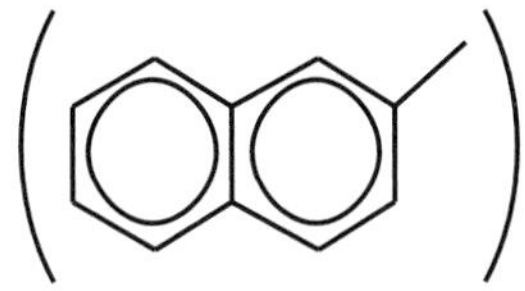

I, CF_3SCN, $C_4F_4H_3$, $CH_2{=}CHCF_2CF_2$-, C_6H_5-C(CN)=C=,

$C_8H_{15}O$, C_9H_{19}, $C_6H_5CF_2$-, R-CH_2CH_2 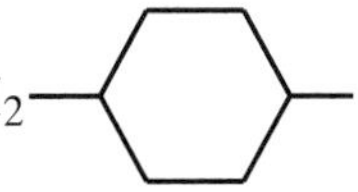 OH,

$C(O)OCH_2CF_3$, $C_7H_{15}N_2$, 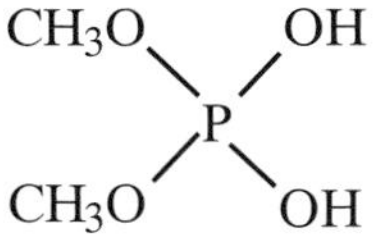

M − 128 Possible Precursor Compounds

Loss of HI from iodides

128 Structural Significance

Quinolines
Naphthalene

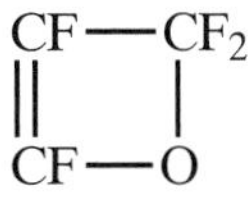

$-CH_2CH_2C(O)N(C_2H_5)_2$, C_3F_4O, $-CF_2CF_2C(O)-$
C_3F_3Cl, HI, $C_2N_2F_4$, $C_3H_3Cl_2F$, $ClC_6H_4O + H$,
$-CH_2CH(CH_3)CH_2N(CH_3)C_3H_7$
$-CH_2CH(CH_3)CH_2N(CH_3)C(O)CH_3$

M − 129 Possible Precursor Compounds

$(CH_3)_3Si\overset{+}{O}{=}CH{-}CH{=}CH_2$ (from the A ring in TMS derivatives of steroids)

129 Structural Significance

Adipates:

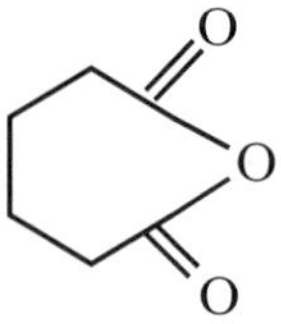

$(CH_3)_3Si-O-CH-CH=CH_2$
C_2FCl_2O, C_6F_3, C_2Cl_3, $C_3H_4FCl_2$, $C_6H_5CH=CHCN$,
$-CF_2Br$, $C_{10}H_9$, $CF_3CH=CCl-$, $H_2N(CH_2)_6NHCH_2-$
$CH_3OC(O)C(CH_3)_2C(O)$,
$-CH_2C(O)-CH_2CH_2C(O)-OCH_3$

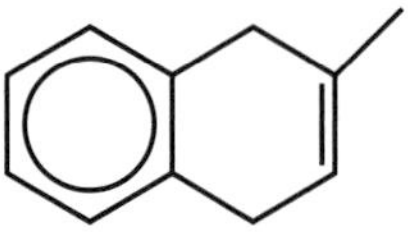

M − 130 Possible Precursor Compounds

130 Structural Significance

Indoles

C_6H_4FCl, $(CH_3)_2CH\text{-}CH=C(OH)OC_2H_5$, $C_6H_{10}O_3$, C_9H_8N,

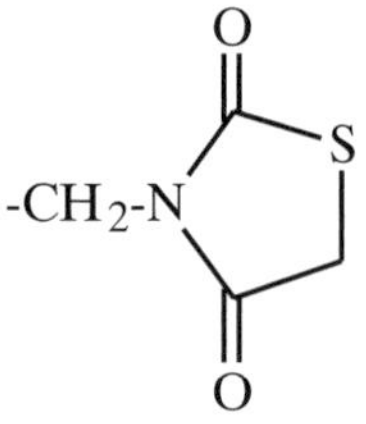

$-CF_2-CF=NCl$,

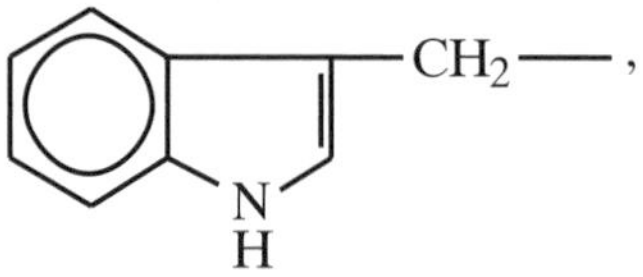

$(CH_3)_2NCH=COC_2H_5$

M − 131   Possible Precursor Compounds

131   Structural Significance

Cinnamates ($C_6H_5CH{=}CHC{\equiv}O$)
Fluorocarbons (C_3F_5)
$-CF_2-CF{=}CF_2$, $CF_3CF{=}CF-$, $CF_3C{=}CF_2$,
  $CHF_2CF_2OCH_2-$, $C_6H_2F_3$, $CH_3OCF_2CF_2-$

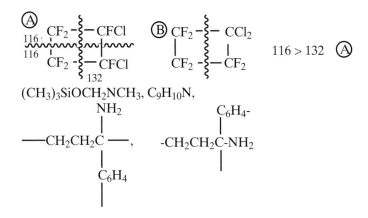

M − 132   Possible Precursor Compounds

132   Structural Significance

Benzimidazoles
$C_6H_3F_3$, $CF_2{=}CCl_2$

$CH_3P(O)Cl_2$

M − 133   Possible Precursor Compounds

$CH_3SCH{-}C(O)OC_2H_5$

133   Structural Significance

Acetanilides
$CH_3SCHC(O)OC_2H_5$, $CH_3C_6H_4C(CH_3)_2$

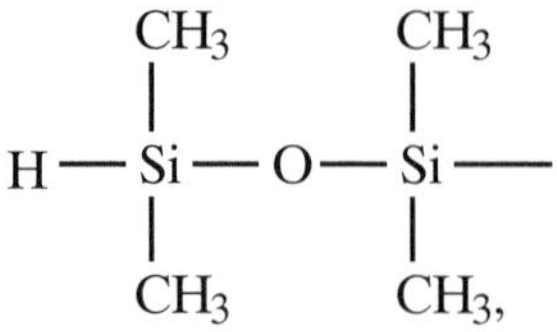

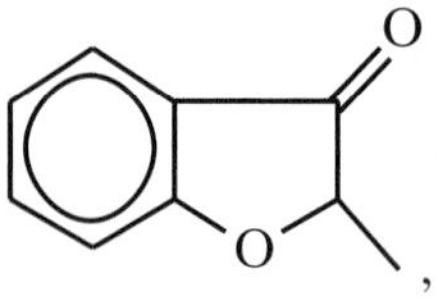

$CF_3CH_2CF_2$-,

$CF_3CF_2CH_2$-, CHF_2CHFCF_2-, $HC \equiv C$—

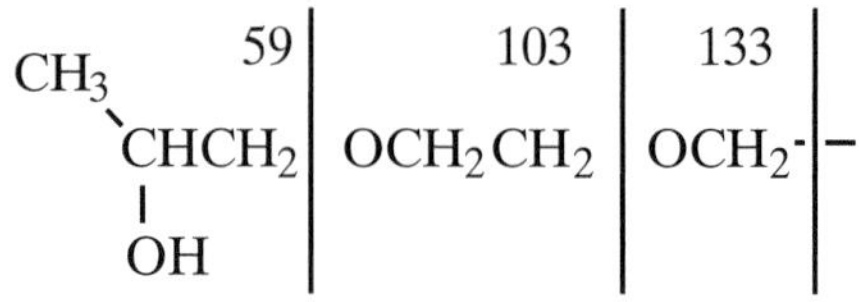

| CH_3 $\diagdown$ $CHCH_2$ | 59 | OCH_2CH_2 | 103 | OCH_2· | 133 |
| OH | | | | | |

$SiCl_3$—

M − 134   Possible Precursor Compounds

134   Structural Significance

$CH_3C_5H_4Mn$—
$CH_3C(CF_2)CH_2NHC(O)$, $-CF_2NHCF_3$
$C_6H_5C(O)NHCH_2$

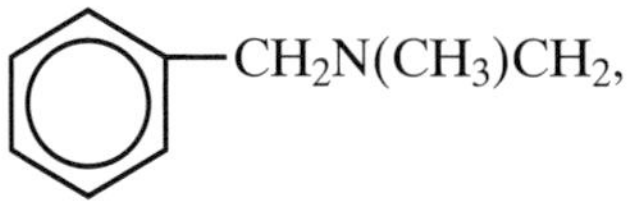

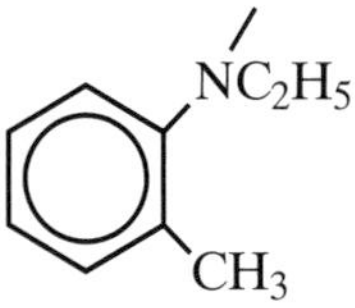

M − 135   Possible Precursor Compounds

135   Structural Significance

Adamantanes ($C_{10}H_{15}$), C_6-C_{18} *n*-alkylbromides $(CH_2)_4Br$
    (with an isotope peak at *m/z* 137)
$C_6H_5C(OH)C_2H_5$, $C_6H_5C(O)OCH_2$–

PTH-amino acids, CH_3O——$C(O)$ -, $-CF_2CF_2Cl$, CF_3CFCl-, CF_3CF_2O-, CF_3OCF_2-, C_8H_7S, $C_6H_5Si(CH_3)_2$,

$C_6H_5Si(H)_2CH_2CH_2$-, —$N=C=S$,

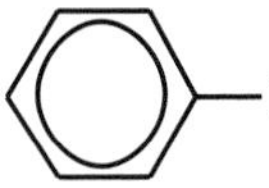

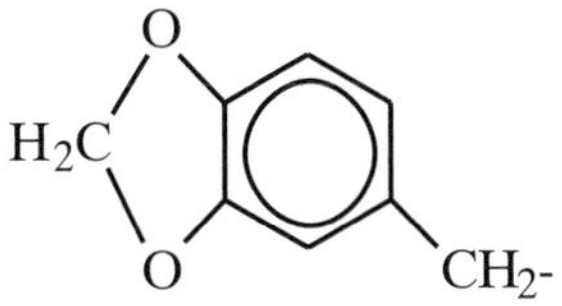

$[(CH_3)_2N]_2P(O)$

M − 136   Possible Precursor Compounds

136   Structural Significance

$O_2N(C_6H_5)CH_2-$, C_5F_4, $CH_2N(CH_2CH_2CN)_2$ (also should observe *m/z* 54)

M − 137   Possible Precursor Compounds

$C_9H_{13}O$

137   Structural Significance

Decalins

C_5F_4H, -$(CH_2)_3SO_3CH_3$, $(C_2H_5O)_2P(O)$

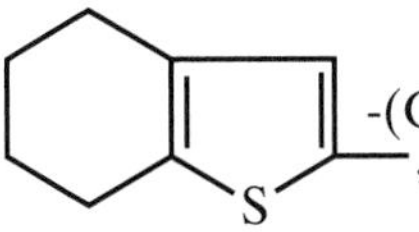

M − 138   Possible Precursor Compounds

138   Structural Significance

Salicylate and [M+H]$^+$ m/z 139

Nitriles [(CH$_2$)$_8$CN] (also may see ions at m/z 96, 110, 124, etc.)

Base peak in dicyclohexylamine and nitroanilines

C$_5$H$_5$FeOH, C$_2$F$_6$, (CH$_2$=CHCH$_2$)$_2$Fe (m/z 110, 124, 138, and 152 suggest nitriles; may observe M − 1 instead of M$^{+\cdot}$)

M − 139   Possible Precursor Compounds

139   Structural Significance

Salicylates (protonated salicylic acid) also may have a large m/z 138 peak, *N*-TFA, *n*-butyl serine

C$_5$F$_4$H$_3$, CF$_3$CH(CH$_3$)CH$_2$C(O), CF$_2$CF$_2$C(CH$_3$)≡C−

M − 140   Possible Precursor Compounds

140   Structural Significance

Methyl esters of dibasic carboxylic acids >dimethyl suberate (usually have a series of peaks at m/zs 84, 98, 112, 126, 140, etc.)

N-TFA butyl ester of alanine

C_4F_4Cl, C_3Cl_3,

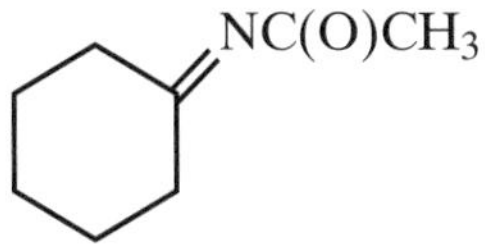

NC(O)CH$_3$

M − 141 Possible Precursor Compounds

141 Structural Significance

Naphthalenes [$C_{11}H_9$ (alkyl naphthalenes should also have *m/z* 91)]

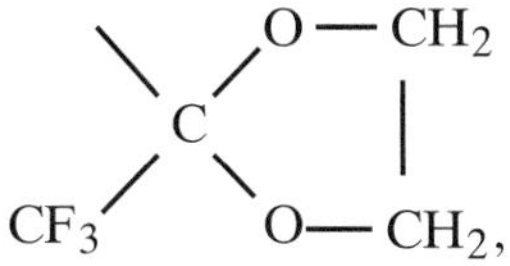

C_6H_5SS-, $-CH_2-$

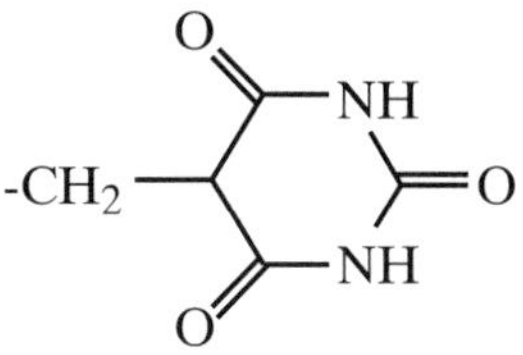

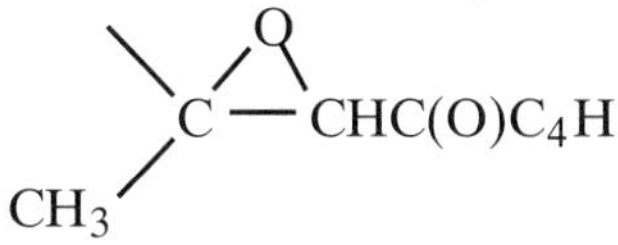

CF$_3$ O—CH$_2$,

$-$CHC(O)C$_4$H$_9$, C_7F_3, $C_4H_7Cl_2O$,

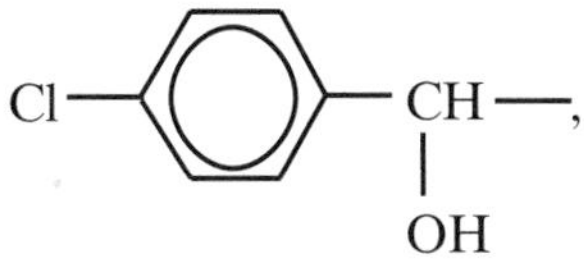

CH$_3$

Cl—⟨ ⟩—CH—, $C_8H_{17}N_2$,

OH

$C_6H_5SO_2$ (also should observe *m/z* 125)

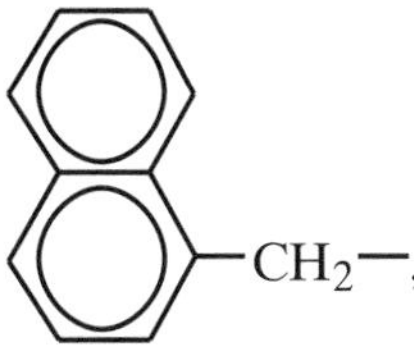

—CH$_2$—, $-CF_2CF=CHC(O)F$, $-CF_2CF=CHCH_2SH$

M − 142 Possible Precursor Compounds

142 Structural Significance

$(C_4H_9)_2NCH_2-$
$C_{10}H_8N$ (from quinolines)
$-CF_2CF_2N_3$

M − 143 Possible Precursor Compounds

143   Structural Significance

Indole and indoline alkaloids
$-C(O)(CH_2)_4CO_2CH_3$ from dimethyl adipate
$(CH_3OC(O)C=CHC(O)CH_3)$

C_4F_5, $C_8H_{15}O_2$, $-C(Cl)=C(Cl)CH_2Cl$, $-CH=C(Cl)CHCl_2$,
$-C(Cl)=CHCHCl_2$, $(C_4H_9)_2SiH$,

$$CH_2=CCH_2OCH_2-$$
with $CO_2C_2H_5$ substituent

M − 144   Possible Precursor Compounds

144   Structural Significance

Indoles

$(CH_3)_2CHCO_2C_3H_7$
with $N(CH_3)_2$ substituent

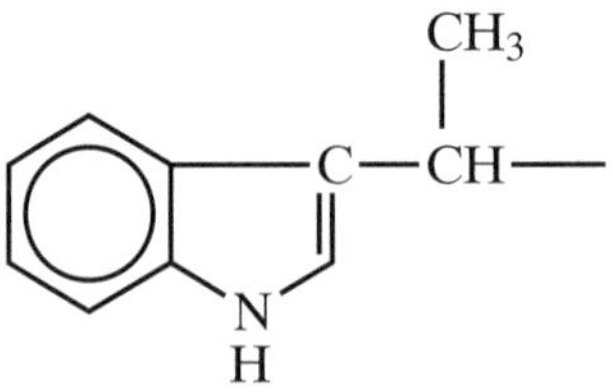

M − 145   Possible Precursor Compounds

The loss of CF_3CF_2CN has been seen from a compound whose probable structure is $(CF_3)_2C=CFC(CF_3)=CFN(CH_3)_2$.

145   Structural Significance

Tetralins
$CF_3OCH_2CF_2-$

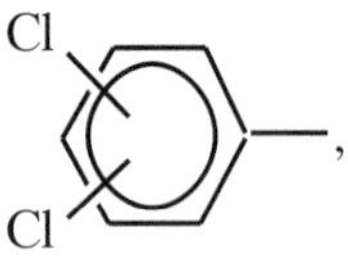

, $CF_3C=CH_2$ (with CF_2), $CFClBr$, $CF_3C_6H_4-$,

$CF_3CH=CHCF_2-$, $C_4F_5H_2$, 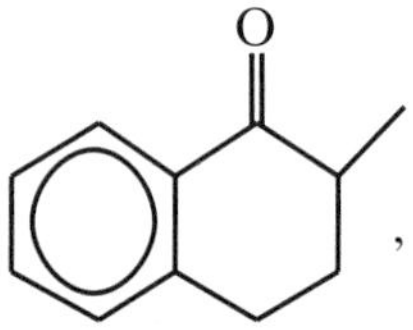 ,

$CCl_3CH_2CH_2-$, $-CCl_2CH_2CH_2Cl$, $(CH_3)_2CHCH\text{-}CHOH\text{-}$
$$\underset{|}{}$$
CO_2CH_3,

$$(CH_3)_2CH-\underset{\underset{CO_2CH_3}{|}}{\overset{\overset{OH}{|}}{C}}-CH_2-$$

M − 146   Possible Precursor Compounds

146   Structural Significance

$CF_3CF_2C{=}NH$

$CFClBrH$, $C_6H_5CF_3$, isomers of

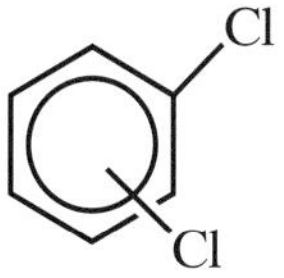

M − 147   Possible Precursor Compounds

147   Structural Significance

$CF_3C(O)CF_2-$, C_3F_5O, $CF_3CCl{=}CF-$, C_3F_4Cl

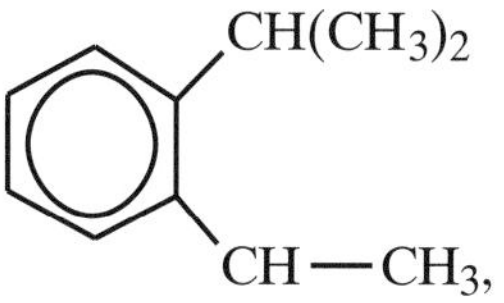

$(CH_3)_2SiOSi(CH_3)_3$, $C_5H_{11}OSiO_2$

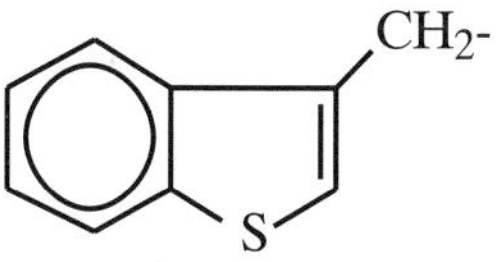

$C_{10}H_{11}O$

M − 148   Possible Precursor Compounds

148   Structural Significance

Aminopurines

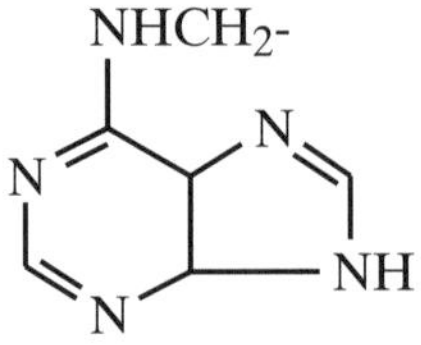

NHCH$_2$-

C_6F_4, C_2FCl_3, $C_6H_3F_2Cl$,

$C_6H_5CH_2CHN(CH_3)_2$, $(CH_3)_2N$—⬡—$C(O)$-

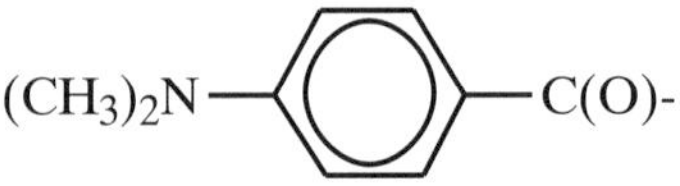

M − 149   Possible Precursor Compounds

149   Structural Significance

Esters of phthalic acid*

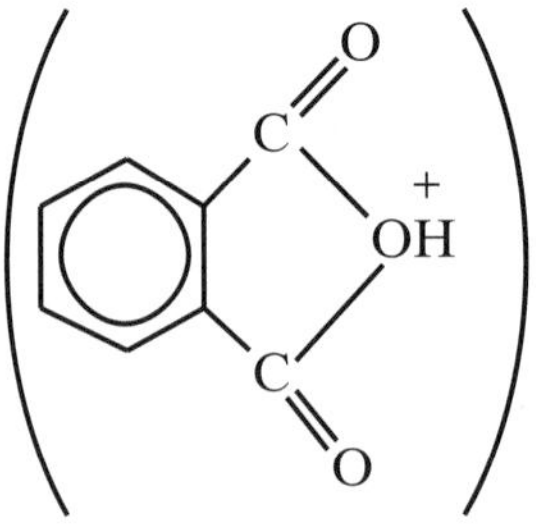

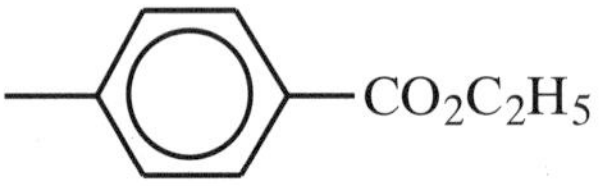

—⬡—$CO_2C_2H_5$

$CF_3CH_2OCF_2$-, $C_2H_5C_6H_4C(O)$-, C_6HF_4, -CF_2SO_3F,

$CFCl_2CHCl$-, ⬡—$CHOC_3H_7$, -$CF_2CH_2CF_2Cl$,

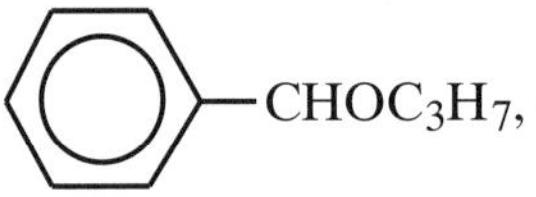

$\overset{\displaystyle C_3H_7}{\underset{\displaystyle C_3H_7}{-Si-}}$Cl, C_2H_5NH—⬡—$NHCH_2$—

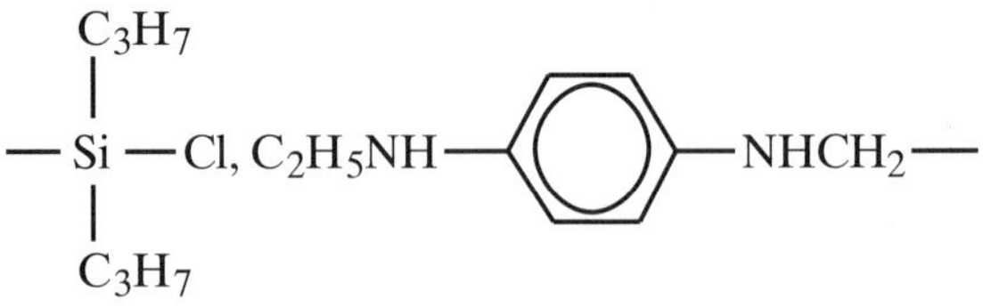

$C_9H_{13}N_2$

*Masses 149, 167, and 279 suggest dioctyl phthalate or di(2-ethylhexyl) phthalate.

M − 150   Possible Precursor Compounds

150   Structural Significance

Steroid alkaloids
$C_3F_6H_2F_4$, $CF_3CF=CF_2$, C_3F_6

M − 151   Possible Precursor Compounds

Loss of $CF_3CFClO-$

151   Structural Significance

$C_9H_{15}C(O)$, $C_5H_{11}CH=CH-CH=CH-C(O)$

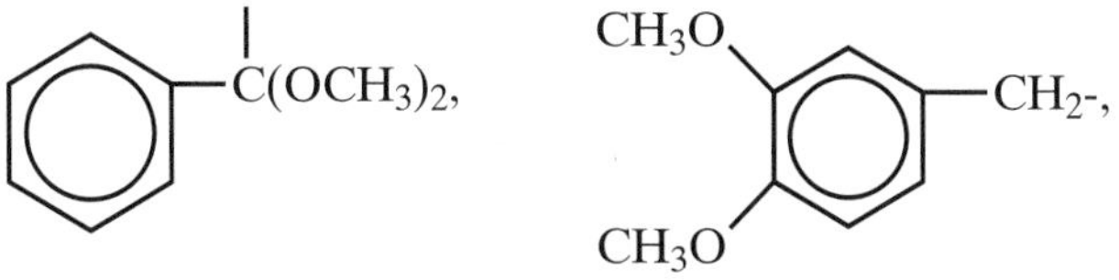

$(CF_3)_2CH-$, CF_3CCl_2-, $CFCl_2CF_2-$, $CF_2ClCFCl-$,
$C_2F_3Cl_2$, $C_6H_3F_4$

M − 152   Possible Precursor Compounds

152   Structural Significance

$(CF_3)_2N-$, C_2F_5NF-,

M − 153   Possible Precursor Compounds

N-TFA, *n*-butylthreonine

153   Structural Significance

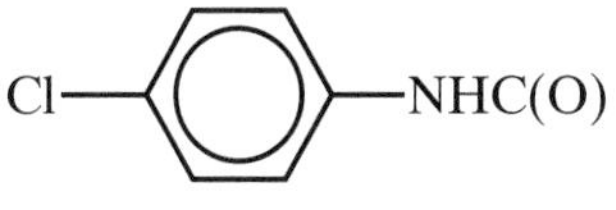

$Cl-$⟨phenyl⟩$-NHC(O)$ (Formation of isocyanate gives *m/z* 153)

M − 154   Possible Precursor Compounds

154   Structural Significance

$$CF_3C(O)NHC(CH_3)_2$$
$$\underset{O-}{|}$$

M − 155   Possible Precursor Compounds

Loss of $C_8H_{17} + C_3H_6$ from steriods (alkyl chain at C17)

155   Structural Significance

$CF_3C(O)OC(CH_3)_2-$, C_5F_5, $CH_3C_6H_4SO_2-$, CH_2CH_2I,
$C_4H_9SiCl_2-$

M − 156   Possible Precursor Compounds

156   Structural Significance

$(C_4H_9)_2NC(O)$, $C_8H_{17}NCH_2$-, CH_3- [piperidine structure]

C_5F_5H, $C_9H_{18}NO$

M − 157   Possible Precursor Compounds

157   Structural Significance

Cl—⬡—S-CH$_2$-,   HO—⬡—SO$_2$-,

(CH$_3$C(O)O)$_2$CHCH=CH-, C$_5$F$_5$H$_2$, ClC$_6$H$_4$NO$_2$

M − 158   Possible Precursor Compounds

158   Structural Significance

C$_{12}$H$_{14}$
CH$_2$C(O)CH$_2$CH$_2$CH$_2$CH$_2$C(O)OCH$_3$
Trichlorobutenes, C$_4$H$_5$Cl$_3$

M − 159   Possible Precursor Compounds

The loss of -C(O)OSi(CH$_3$)$_2$C(CH$_3$)$_3$ is characteristic of TBDMS derivatives of amino acids

159   Structural Significance

C$_6$H$_5$CCl$_2$-,   Cl(⬡)—CH$_2$-
                      Cl

C$_6$H$_5$SCF$_2$-, (C$_4$H$_9$O)$_2$CH-,   CF$_3$\C//N
                                          CF$_2$/ ‖N,

                            CF$_3$
                            |
-CF$_2$CF$_2$C(O)OCH$_3$, F—⬡—SO$_2$-, −CFC(O)OCH$_3$

M − 160   Possible Precursor Compounds

160   Structural Significance

CF_3—⟨ ⟩—NH, $[(CH_3)_2N]_3Si$-,

M − 161   Possible Precursor Compounds

161   Structural Significance

CCl_2Br—, $CHF_2CF_2OCH_2OCH_2$—

M − 162   Possible Precursor Compounds

162   Structural Significance

$(C_2H_5)_2N$—⟨ ⟩—, C_4F_6,

M − 163   Possible Precursor Compounds

163   Structural Significance

$CF_3CF_2CH_2OCH_2-$, $CHF_2CHFCF_2OCH_2-$
$CF_3CH_2CF_2OCH_2-$, $C_3F_3Cl_2$, C_4F_6H, C_3F_4ClO,

$CH_3OC(O)$ ⟨benzene ring⟩ $C(O)-$

M − 164   Possible Precursor Compounds

164   Structural Significance

CCl_3CCl-, $-CF_2N=CFCF_3$, $C_4F_6H_2$, $C_6H_3FCl_2$, $C_7H_{10}Cl_2$

M − 165   Possible Precursor Compounds

165   Structural Significance

Dinitrotoluenes

$C_{13}H_9$:

CCl_3CHCl-, $Sn(CH_3)_3$
(Sn isotope peaks at *m/z* 161, 163, 165)

M − 166   Possible Precursor Compounds

166 Structural Significance

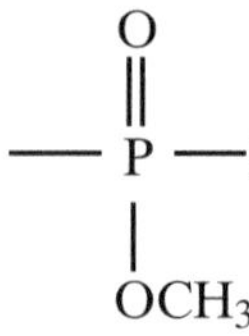 $NHSi(CH_3)_3$, $C_6H_2F_3Cl$, 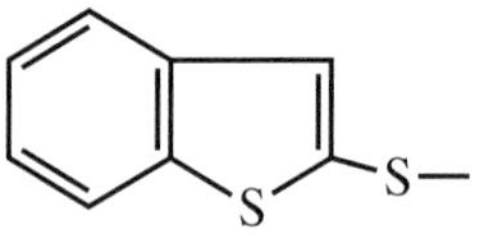

M − 167 Possible Precursor Compounds

167 Structural Significance

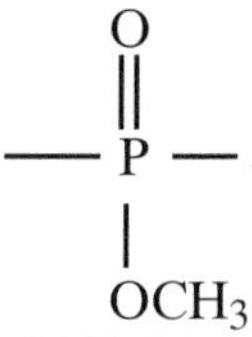 $Si(CH_3)_3$, $CF_3CFHOCF_2$-, CCl_3CF_2-, C_6F_5-,

$CFCl_2CFCl$-, CF_2ClCCl_2-, $(C_6H_5)_2CH$, $C_2F_2Cl_3$

Certain phthalates, $(CH_3)_3C$ 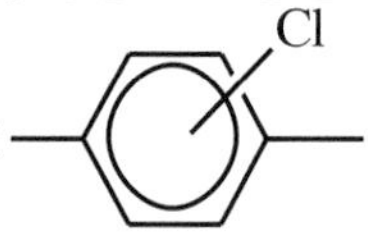

M − 168 Possible Precursor Compounds

168 Structural Significance

Dichlorophenyl phenyl ethers
$CF_3C(O)NH(CH_2)_4$—, $C_6H_9F_3ON$, C_6F_5H

M − 169 Possible Precursor Compounds

Loss of C_3F_7 from perfluoro compounds

169 Structural Significance

$CH_3C(O)CH_2$ 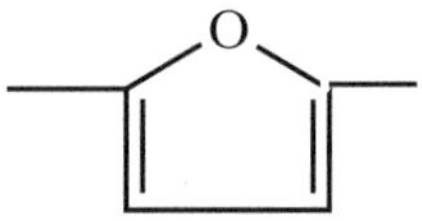 CHOH, $CF_3CF_2CF_2$-,

$C_6F_5H_2$, $C_{10}H_{21}C(O)$

M − 170 Possible Precursor Compounds

170 Structural Significance

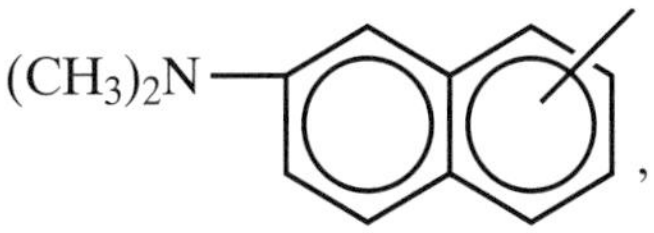
(CH3)2N

, $C_2F_4Cl_2$,

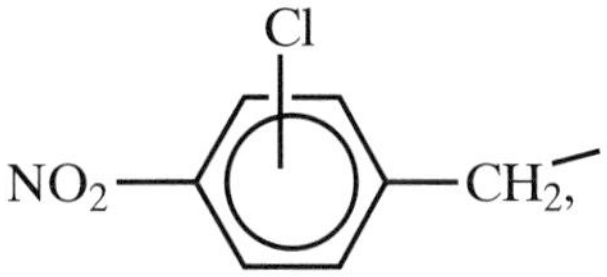
Cl
NO2
CH2,

$(C_5H_{11})_2NCH_2$-

M − 171 Possible Precursor Compounds

Loss of

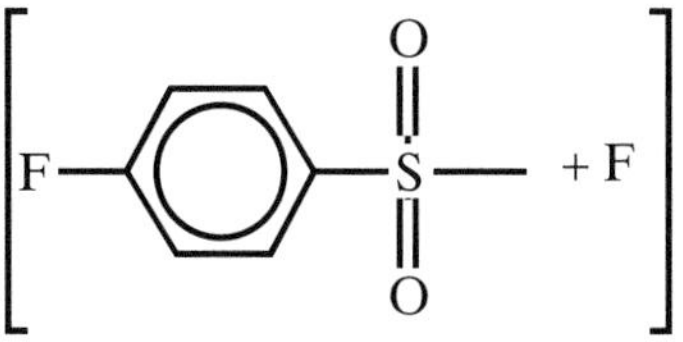
F
O
S
O
+ F

171 Structural Significance

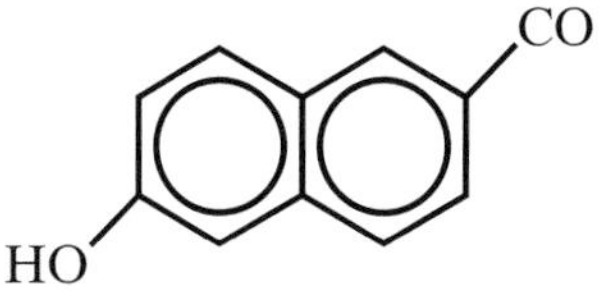
CO
HO

M − 172 Possible Precursor Compounds

172 Structural Significance

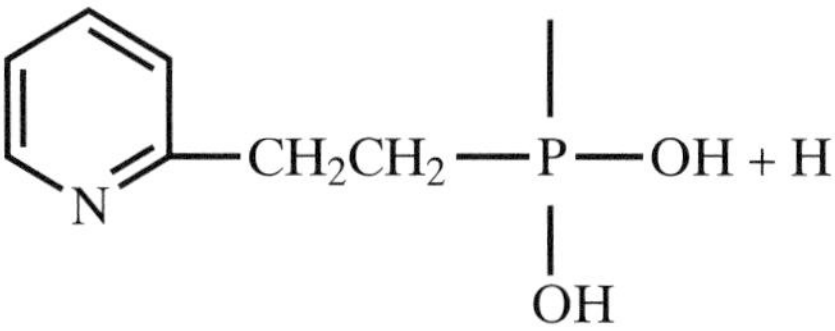
N
CH2CH2
P
OH + H
OH

M − 173 Possible Precursor Compounds

173   Structural Significance

$$C_8H_{17}-\underset{\underset{}{\overset{\overset{CH_3}{|}}{Si}}}{\overset{}{}}-OH,$$

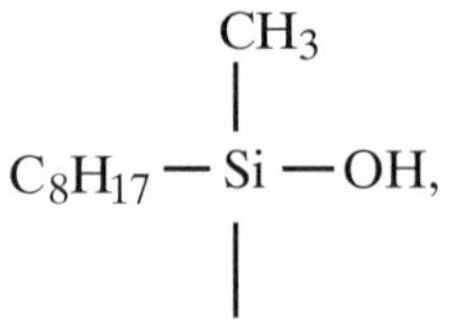

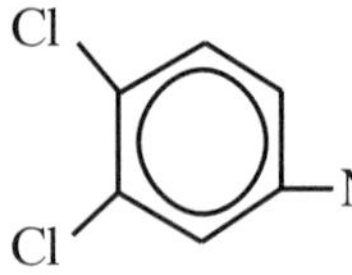

$-N=N-, CF_3C_6H_4C(O),$

$-CHC(O)OC_2H_5, C_8H_5F_3O, CF_3C_6H_4N=N-$
$\quad|$
$\;CH_2CO_2C_2H_5$

M − 174   Possible Precursor Compounds

174   Structural Significance

$-CH_2N(SiC_3H_9)_2$* loss is observed when

$-\underset{\underset{OH}{|}}{CHCH_2N(TMS)_2}$

or $-(CH_2)_3N(TMS)_2$ is present

M − 175   Possible Precursor Compounds

175   Structural Significance

$C_5F_6H,$

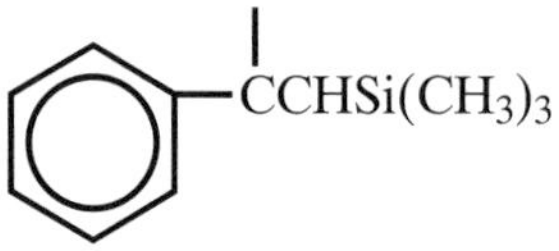

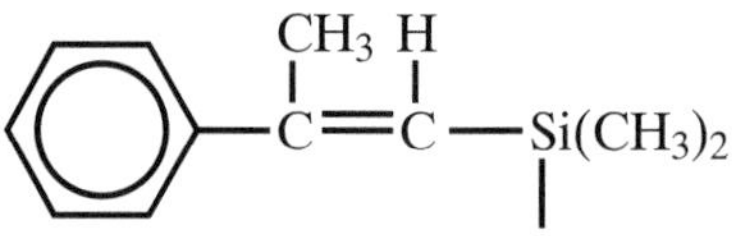

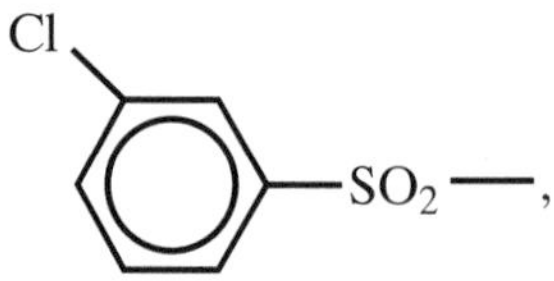

$-\overset{\overset{|}{}}{C}CHSi(CH_3)_3$

$-\underset{\underset{Si(CH_3)_2}{|}}{C}=\overset{\overset{H}{|}}{C}-Si(CH_3)_2$

with CH_3 on the first carbon

$-SO_2-,\quad CH_3CO_2CH_2\underset{\underset{CH_2-}{|}}{\overset{\overset{OH}{|}}{C}}CO_2CH_3$

*Apparently two TMS groups can add to the amino group, especially for $-(CH_2)_nN$, when $n > 2$.

M − 176   Possible Precursor Compounds

176   Structural Significance

$$\overset{\displaystyle CN}{\underset{\displaystyle }{|}}$$
$CF_3CF\text{-}CF_2\text{-}$, $C_5F_6H_2$

M − 177   Possible Precursor Compounds

177   Structural Significance

SnC_4H_9 (*m/z* 177, 175, 173)

$C_6H_5NHC(O)CH_2C(O)CH_3$, $C_5F_6H_3$, CF_3 $\diagdown$ $\diagup$ $\overset{O-CHF}{\underset{O-CHF}{\underset{|}{C}}}$

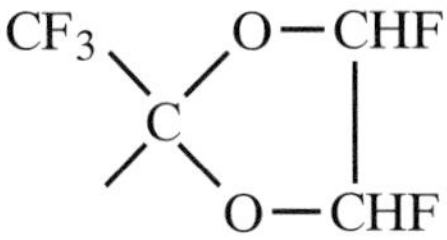

$(C_4H_2F_5O_2)$ $\quad$ $CO_2C_2H_5$

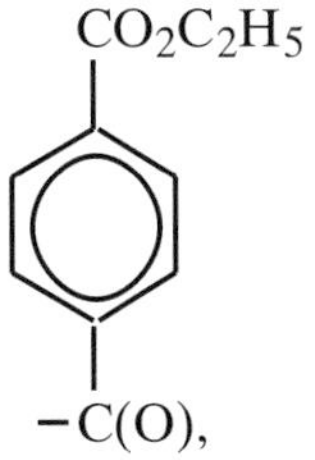

$(C_4H_9)_2SiCl$

$-C(O)$,

M − 178   Possible Precursor Compounds

178   Structural Significance

C_4F_5Cl, $CF_2{=}CFCF{=}CFCl$

M − 179   Possible Precursor Compounds

179   Structural Significance

$C_6H_5CHOSi(CH_3)_3$, $C_3F_3Cl_2O$, $(CF_3)_2CHC(O)-$, C_4HF_6O
$(CF_3)_2C{=}COH$
$C_{10}H_{11}O_3$

M − 180   Possible Precursor Compounds

180   Structural Significance

$C_6H_3Cl_3$
$C_6H_5CH_2OC(O)CO_2H$

181   Structural Significance

$$\underset{\displaystyle C_4F_7,\ \ CF_3\overset{\displaystyle CN}{\underset{|}{C}}F\text{-}CF_2\text{-},\ C_5F_6H_2}{}$$

M − 182   Possible Precursor Compounds

182   Structural Significance

$(C_6H_5)_2Si-$ (see tetraphenylsilane)

M − 183   Possible Precursor Compounds

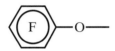

183   Structural Significance

$C_2FCl_4,\ \ -CF_2CF_2SO_2F,\ \ -CFCl_2CCl_2-,\ \ (C_6H_5)_2\overset{|}{C}OH,$
$(C_6H_5)_2SiH$

M − 184   Possible Precursor Compounds

184   Structural Significance

$$CH_3C(O)N\overset{\displaystyle \diagup C_8H_{17}}{\diagdown CH_2-}$$

M-185   Possible Precursor Compounds

185   Structural Significance

$C_8H_{11}NO_2S$
Tributyl citrate (MW = 360)

$$\text{(pyridyl-phenyl)}-OCH_2-\ ,\ (C_2H_5O)_2PS_2,$$

$(CH_3)_2SnCl$, (*m/z*s 185, 183, and 181; isotopes for Sn)

-CH$_2$N $\diagup\diagdown$ N-CH$_2$CH$_2$OC(O)CH$_3$, $(C_6H_5)_2CF$-,

$C_9H_{19}N_3O$,

$C_{11}H_{21}O_2$, $C_{10}H_{17}O_3$, $C_4H_9OC(O)(CH_2)_4C(O)$—
(*m/z*s 185 and 129 suggest dibutyl adipate)

M − 186   Possible Precursor Compounds

186   Structural Significance

C_6F_6

M − 187   Possible Precursor Compounds

187   Structural Significance

$C_{13}H_{15}O$, C_6F_6H

M − 188   Possible Precursor Compounds

188   Structural Significance

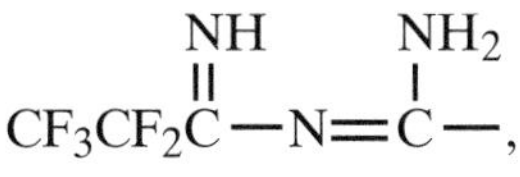

$$CF_3CF_2\overset{\overset{\textstyle NH}{\|}}{C}-N=\overset{\overset{\textstyle NH_2}{|}}{C}-, \quad C_6F_6H_2$$

M − 189   Possible Precursor Compounds

189   Structural Significance

$$CF_3CO_2\overset{|}{C}(CH_3)CH_2Cl, \quad CH_3-\overset{\overset{\textstyle CN}{|}}{\underset{\underset{\textstyle H}{|}}{C}}-NHC(O)NH$$

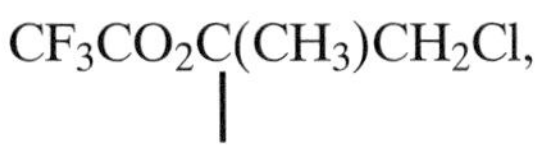

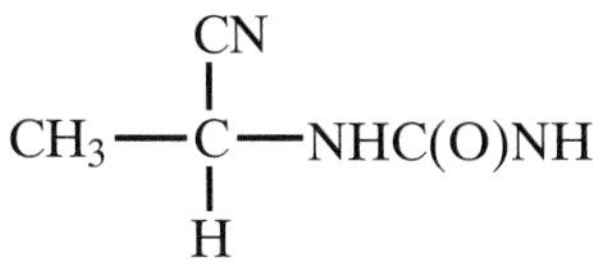

M − 191   Possible Precursor Compounds

191 Structural Significance

m/z 191, 204, and 217 suggest a TMS hexose
$(CH[OSi(CH_3)_3]_2)$

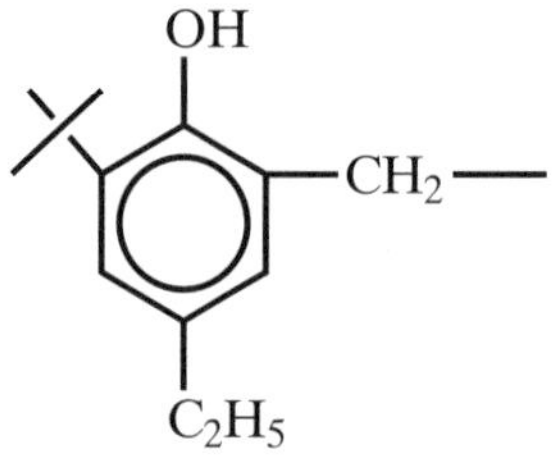

OH

X —— CH_2 ——

C_2H_5

M − 193 Possible Precursor Compounds

193 Structural Significance

$-SnSi(CH_3)_3$ (m/z 193, 191, 189)
C_5F_7, $(C_2H_5O)_3SiOCH_2-$

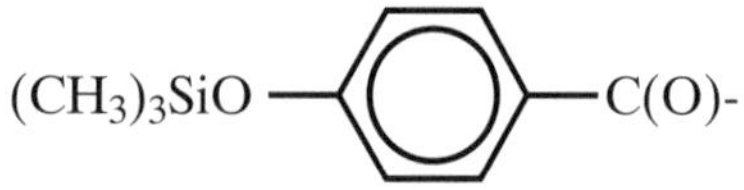

$(CH_3)_3SiO$ ——〈 〉—— $C(O)-$

M − 195 Possible Precursor Compounds

195 Structural Significance

$C_3F_2Cl_3O$, $C_5F_7H_2$, CF_3 —〈 〉— CF_2 ——

CH_2CH_2-

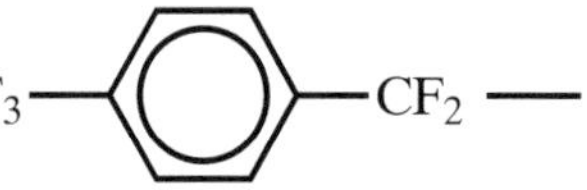

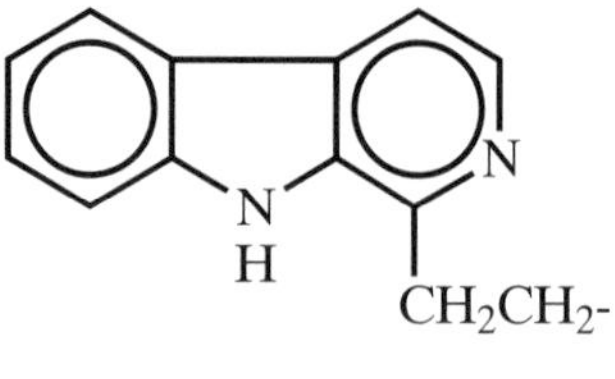

M − 196 Possible Precursor Compounds

196   Structural Significance

$C_5F_7H_3,$

M − 197   Possible Precursor Compounds

197   Structural Significance

$C_4F_7O,$ $-C(O)-, \ C_{13}H_9O_2$

M − 198   Possible Precursor Compounds

Loss of two CD_3SiOH groups from deuterated TMS derivative (e.g., cholic acid)

198   Structural Significance

M − 199   Possible Precursor Compounds

199   Structural Significance

$(C_4H_9)_3Si-$
$CF_3CF_2CF_2OCH_2-$

M − 200   Possible Precursor Compounds

200   Structural Significance
          C_4F_8

Part III References

Additional Correlations

Gudzinowicz, B. J., Gudzinowicz, M. J., and Martin, H. F. *Fundamentals of Integrated GC-MS, Part II: Mass Spectrometry* (Vol. 7, Chromatographic Science Series). New York: Marcel Dekker, 1976.

McLafferty, F. W., and Venkataraghaven, R. *Mass Spectral Correlations*, 2nd ed. (Advances in Chemistry Series). 1982.

Part IV

Appendices

Appendix 1

Definitions of Terms Related to Gas Chromatography

Adjusted retention time (t'_R): The retention time for a substance (t_R) minus that of an unretained substance (t_m): $t'_R = t_R - t_m$.

Adjusted retention volume (V'_R): The retention volume for a substance (V_R) minus the retention volume of an unretained substance (V_m): $V_R' = V_R - V_m$.

Bleeding: An appearance of a background signal from a chromatographic system, caused by the stationary phase or contamination of the inlet system. The column bleed usually increases with increasing column temperature.

Capillary column: A narrow bore tube (0.25–1 mm ID) typically 30–100 m long (usually of deactivated fused silica), whose walls are coated with a liquid stationary phase to produce high-efficiency separations ($N > 100,000$).

Conditioning: Equilibrating a column with a flow of carrier gas (mobile phase) at the maximum expected operating temperature of the column.

Dead volume: The total volume in the system that is swept by the carrier gas.

Derivitization: Chemical reaction of a sample that yields a product that is more volatile and stable and that has improved gas chromatographic behavior over the original substance. See Appendix 3 for derivatives found useful in GC/MS.

Efficiency: Degree of band broadening for a given retention time. It is expressed as the number of theoretical plates, N, or as the height equivalent to a theoretical plate, *HEPT*.

Gas holdup: V_m is the volume of carrier gas that passes through the column to elute an unretained substance, such as argon or methane. The time required is t_m.

Height equivalent to a theoretical plate (*HETP*): A measure of the efficiency of a column usually expressed in millimeters. $HETP = L/N$, where L is the length of a column and N is the number of theoretical plates. The reciprocal of *HETP* is also used to describe efficiency and is often expressed by the terms *plates per meter*.

Resolution: The degree to which two peaks are separated. This is a function of the number of theoretical plates, N, in a column and the separation factor between the two components.

Retention time (t_R): The time required for a substance to pass through the column and be detected.

Retention volume (V_R): The volume of carrier gas that passes through the column to elute a substance.

Selectivity: The characteristics of the stationary phase that determine how far apart the peak maxima of two components will be separated.

Theoretical plates (N): A measure of the efficiency of the column or sharpness of the peaks. The more theoretical plates a column has the narrower the peaks will be:

$$N = 16 \left(\frac{t_R}{W} \right)^2$$

where W is the width of the peak at baseline.

Van Deemter equation: An equation relating efficiency (*HEPT* in mm) to linear flow velocity in a chromatographic column. The efficiency is expressed as the height equivalent to a theoretical plate ($HEPT = A + B/V + Cv$), where A, B, and Cv are constants and V is the linear velocity of the carrier gas. This equation tells us that to obtain maximum efficiency, the carrier gas flow must be optimized.

For additional terms consult:
Grob, R. L., ed. *Modern Practices of Gas Chromatography*. New York: Wiley-Interscience, 1985.

Appendix 2

Tips for Gas Chromatography

Because oxygen damage to a stationary phase is one of the most common reasons that a capillary column fails, the carrier gas should be cleaned by passing it through an indicating moisture trap followed by an indicating oxygen trap. These traps remove oxygen, water, hydrocarbons, sulfur, and chlorine compounds.

Number of units required	Description	Chrompack* cat. no.
1	Oxygen trap	7970
1	Moisture trap	7971
2	Connecting units	7980 (1/4″ fittings) or 7988 (1/8″ fittings)
2	Wall-mounting brackets	7981

*Or equivalent.

Always adjust the carrier gas linear velocity to 20–40 cm/sec before heating the capillary column. Neglecting to supply carrier gas while heating a capillary column will destroy it. Avoid injecting samples containing inorganic bases or acids into a capillary column because they will strip the liquid phase off the column.

Hints for Improving Capillary GC Resolution

Resolution in GC is dependent on the stationary phase, the stationary phase thickness, column length, column diameter, column temperature, and the linear velocity of the carrier gas.[1,2]

Stationary Phase Selection

For routine separations, there are about a dozen useful phases for capillary columns. The best general-purpose columns are the dimethylpolysiloxane (DB-1 or equivalent) and the 5% phenyl, 95% dimethylpolysiloxane (DB-5 or equivalent). These relatively nonpolar columns are recommended because they provide adequate resolution and are less prone to bleed than the more polar phases. If a DB-1, DB-5, or equivalent capillary column does not give the necessary resolution, try a more polar phase such as DB-23, CP-Sil88, or Carbowax 20M, providing the maximum operating temperature of the column is high enough for the sample of interest. See Appendix 3 for fused silica capillary columns from various suppliers.

If a DB-1 or DB-5 capillary column does not separate the components, make use of the sample functionality stationary phase interactions presented by McReynolds[3] (Table A.1).

Table A.1. Interaction of liquid phase with certain functionalities*,†

	X'	Y'	Z'	U'	S'
Interaction	Intermolecular forces	Electron attractors	Electron repellers	Complex	Complex
Compound types	Aromatics Olefins	Alcohols Nitriles Acids CHCl$_2$ CCl$_3$ CH$_2$Cl NO$_2$ Diols	Ketones Ethers Aldehydes N(Me)$_2$ Esters Epoxides	Nitromethane Nitrogroups	Pyridine Dioxane Aromatic bases

*An example of how the table is used: To separate an alcohol and a ketone with the same boiling point, pick a stationary phase with a high Z' value with respect to the Y' value, then the ketone should elute after the alcohol (e.g., DB-210, where Z' = 358 and Y' = 238).
†See Appendix 5 text for values of X', Y', Z', U', and S'.

Stationary Phase Film Thickness

For most applications, a 1 μm film thickness is preferred because it is more universal, has less absorption, and allows for higher sample load. A thinner (0.33 μm) stationary phase is useful for higher-boiling and heat-sensitive compounds. A thicker stationary phase is better for low-boiling compounds.

Column Length

Thirty-meter columns are typically used. Shorter (5–15 m) columns are advantageous for thermally labile and higher-boiling compounds.

Column Diameter

Most of the GC conditions given in this book are for 0.25-mm ID columns, but 0.32- or 0.53-mm ID columns also can be used. The wide bore fused silica columns are found to be more inert, probably because of the greater film thicknesses. A splitter arrangement with a jet separator is used with 0.53-mm ID columns. This arrangement shown in Figure 11.1 has the advantage of simultaneous flame ionization quantitation.

Column Temperature

The maximum column temperatures used in GC/MS are usually 25–50° lower than those used in capillary GC with a flame ionization detector. Higher temperatures can be used in GC/MS but there will be more column bleed, which will require more frequent cleaning of the ion source of the mass spectrometer.

Every column (including chemically bonded columns) will have some column bleed. The amount of column bleed will increase with increasing column temperature, film thickness, column diameter, and column length. The base line starts to rise approximately 25–50° below the upper temperature limit of the stationary phase. After a column is installed in a GC/MS system, a background spectrum should be obtained for future reference.

Linear Gas Velocity

An important parameter when considering GC resolution of the sample components is the carrier gas linear velocity (flow rate, F), which can be determined by injecting 5–50 μl of argon or butane and measuring the time from injection to detection by the mass spectrometer. An optimum linear velocity using helium as a carrier gas is approximately 30 cm/sec and is determined by dividing the column length in centimeters (L) by the retention time (t_R) of argon or butane.

$$F = \frac{L}{t_R}$$

For example, a 30-m column (regardless of diameter) should have a t_R for argon or butane of approximately 100 sec. It appears better to set the linear velocity higher than the optimum rather than lower than the optimum to obtain good column efficiency. Determine the column temperature where the most difficult-to-separate compounds elute and set the linear velocity at that temperature. Now the column will exhibit its maximum resolving power at the point where it is needed most.

Splitter Vent Flow

Once the linear gas velocity of the carrier gas has been set, the splitter vent flow should be adjusted between 15 and 100 ml/min. Use lower values to improve detection and higher values to improve GC resolution and decrease column overload.

Capillary column ID	0.25	0.32	0.53
Recommended splitter vent flow rate (ml/min)	100	40	20
Maximum amount of each sample component (ng)	100	500	2000

Injector

The injector temperature should be determined by the nature of the sample and the volume injected, not by the column temperature. When analyzing biological or high-boiling samples, clean the injector body with methanol or other suitable solvent once per week. Install a clean packed injector liner and a new septum, preferably near the end of a workday. Program the column to its maximum temperature, then cool the column and run a test mixture to check the system using standard conditions.

If the injection port temperature is too high, the sample will partially decompose but the GC peak for the intact material will be sharp and t_R will not be affected. However, if the sample decomposes on the column the t_R will vary and the observed peak will be broad. For normal stable samples, the injection port temperatures should be in the range of 150–280°.

Transfer Line Temperature

In general, the temperature of the transfer line should be 25° above the maximum column temperature for the sample.

Sample Injection

Contrary to popular opinion, slow injection of a large sample (3–5 μl) by split injection is best. The mass spectrometer total ion current will return to the baseline more quickly using a slow injection technique.

Split Injection

Recommended injection volume with a split injection is 1–2 μl. If a large sample has to be injected, inject it slowly as previously mentioned. Typical split ratios are 1:10 to 1:100. (Use a packed liner or other suitable liner.)

Table A.2. Suggested column temperatures for splitless injection*

Solvent	BP	Suggested initial column temperature
Diethyl ether	36°	10–25°
n-Pentane	36°	10–25°
Methylene chloride	40°	10–30°
Carbon disulfide	46°	10–35°
Chloroform	61°	25–50°
Methanol	65°	35-55°
Tetrahydrofuran (THF)	66°	35-55°
n-Hexane	69°	40-60°
Ethyl acetate	77°	45–65°
Methylethyl ketone	80°	45–65°
Acetonitrile	82°	50–70°
n-Heptane	98°	70–90°
Toluene	111°	80–100°
Dimethylformamide	153°	100–120°
Dimethylacetamide	163–165°	120–140°

*Injection temperature: 150–200°; purge activation times: 25–120 sec.

Splitless Injection (For Trace Analyses)

The purge activation time (or the sample transfer time) depends on the sample solvent and carrier gas flow relative to the volume of the injection port liner and the boiling points of the sample components. For most applications, a purge activation time of 50–120 sec is better than 25–50 sec. Early activation results in the loss of sample, while late activation results in peak tailing. A more accurate method of determining purge activation time is to divide the volume of the injector liner by the flow rate (F) of the carrier gas and multiply this value by 1.5 or 2.0. (Do not use a packed liner.)

After the purge activation time (which can be optimized), the split mode is turned on and the solvent vapors are purged out of the split vent. If possible, choose a sample solvent that has a boiling point that is at least 20° below the boiling point of the first sample component (if known).

The most frequent mistake in splitless injection is that the initial column temperature is too high. We recommend values shown in Table A.2.

References

1. Grob, K. *Classical Split and Splitless Injection in Capillary GC.* Heidelberg, Germany: Huethig Publishing, 1988.
2. Rood, D. *A Practical Guide to the Care, Maintenance and Troubleshooting of Capillary Gas Chromatographic Systems.* Heidelberg, Germany: Huethig Publishing, 1991.
3. McReynolds, W.O. *J. Chromat. Science, 8,* 685–691, 1970.

Appendix 3

Derivatives Found Useful in GC/MS

Functional Group

$-OH$

TMS Derivative	Increase of MW per Hydroxyl
$-O\ Si(CH_3)_3$	72

TBDMS Derivative

$$-O\underset{\underset{CH_3}{|}}{\overset{\overset{CH_3}{|}}{Si}}-C(CH_3)_3$$

114

Acetyl Derivative

$$-O\overset{\overset{O}{\|}}{C}CH_3$$

42

TMS Procedure*

1. Add 100 μl MSTFA reagent (Pierce cat. no. 48911) or 100 μl BSTFA reagent (Pierce cat. no. 38832) to 1–5 mg dry sample.
2. Add 50 μl pyridine or other solvent, such as acetonitrile.
3. Cap the vial and mix well.
4. Heat at 60° for 5 min.
5. Cool the reaction mixture and inject 1–2 μl into the GC/MS system.

TBDMS Procedure

1. Add 100 μl acetonitrile or other suitable solvent to 1–5 mg dry sample.
2. Add 100 μl MTBSTFA reagent (Pierce cat. no. 48920) to the sample and heat at 60° for 15 min.
3. Inject 1–2 μl into the GC/MS system.

Acetyl Procedure

1. Add 40 μl of pyridine and 60 μl acetic anhydride to the dry sample in a septum vial.
2. Heat at 60° for at least 30 min.
3. Evaporate to dryness with clean, dry nitrogen.
4. Dissolve the residue in 25 μl ethyl acetate.
5. Inject 1–2 μl into the GC/MS system.

Functional Group

$-COOH$

Methyl Ester Derivative	**Increase of MW**
$-COOCH_3$	14

TMS Derivative

$$-\overset{O}{\overset{\|}{C}}-O\,Si(CH_3)_3 \qquad 72$$

*Note: TMS derivatives cannot be injected into a DB-Wax column.

TBDMS Derivative

$$- \overset{\overset{\displaystyle O}{\|}}{C} - O \overset{\overset{\displaystyle CH_3}{|}}{\underset{\underset{\displaystyle CH_3}{|}}{Si}} C(CH_3)_3 \qquad\qquad 114$$

Methyl Ester Procedures

A. Anhydrous methanol/H_2SO_4 procedure*
1. Add 250 μl methanol and 50 μl concentrated sulfuric acid to 1–5 mg dry acid sample.
2. Cap the vial and heat at 60° for 45 min.
3. Cool the reaction mixture and add 250 μl distilled water with a syringe.
4. Add 500 μl chloroform or methylene chloride.
5. Shake the mixture for 2 min.
6. Inject 1–2 μl chloroform or methylene chloride (bottom layer) into the GC/MS system.

B. Anhydrous methanolic/HCl procedure
1. Add 250 μl 3N methanolic HCl (Supelco cat. no. 3-3050 or 3-3051) to 1–5 mg dry sample.
2. Cap tightly with a teflon-lined cap and heat at 60° for 20 min.
3. Cool the reaction mixture and neutralize carefully before sampling.†
4. Inject 1–2 μl into the GC/MS system.

C. BF_3/methanol procedure
1. Add 250 μl BF_3/methanol reagent (Pierce cat. no. 49370) to 1–5 mg dry sample.
2. Cap tightly and heat at 60° for 20 min.
3. Cool in an ice-water bath and add 2 ml distilled water.
4. Extract within 5 min with 2 ml methylene chloride.
5. Extract with another 2 ml methylene chloride.
6. Combine the extracts and evaporate the methylene chloride with clean, dry nitrogen to approximately 100 μl, or evaporate to dryness and add 100 μl methylene chloride.
7. Dry the extract by adding a small amount of anhydrous sodium sulfate.

---

*This is an old method, but it is found to be best for trace components. A byproduct is dimethyl sulfate (MW = 126), which shows ions at *m/z*s 95, 96, and 66.
†HCl does not appear to adversely affect the GC columns like other mineral acids.

D. Methyl-8 procedure
1. Add 100 μl Methyl-8 reagent (Pierce cat. no. 49350) to 1–5 mg dry sample.
2. If necessary, add 100 μl pyridine, chloroform, methylene chloride, THF, or DMF.
3. Cap tightly and heat at 60° for 15 min.
4. Inject 1–2 μl into the GC/MS system.
E. Method for determining the number of —COOH groups
To aid in determining the number of carboxyl groups, prepare a derivative using trideuterated Methyl-8 (Pierce cat. no. 49200), using the same procedure previously given. Inject 1–2 μl of the trideuterated methyl ester separately or mix equal portions of the nondeuterated methyl ester with the trideuterated methyl ester, and inject 2 μl immediately into the GC/MS system. From the mass difference, it is easy to determine the number of carboxyl groups present.

TMS Procedure for Acids

A. For low molecular weight aliphatic acids, try DMSDEA reagent (Pierce cat. no. 83010). Otherwise, use MSTFA, BSTFA, or TRI-SIL BSA/ Formula P (Pierce cat. no. 49011). For keto acids, the methoxime derivative should be prepared first. (See the procedure given under Keto Acids, Chapter 4, IIIB).
1. Add 200 μl MSTFA (Pierce cat. no. 48911), BSTFA (Pierce cat. no. 38832), or TRI-SIL/BSA (Pierce cat. no. 49012) to 1–5 mg dry sample.
2. Cap tightly and heat at 60° for 15 min.
3. Inject 1–2 μl into the GC/MS system.
B. Method for determining the number of —COOH groups in a molecule
The TMS derivative of an acid can be converted to the methyl ester using anhydrous methanolic HCl.[1] Obtain a mass spectrum of the TMS derivative of the acid, and then evaporate the TMS reaction mixture with clean, dry nitrogen. Add 250 μl of anhydrous methanolic HCl (Pierce cat. no. 33050) and heat at 60° for 20 min. Many TMS derivatives of acids are converted to methyl esters at room temperature after 20 min. If the sample is rerun as the methyl ester, the number of carboxyl groups can be determined by the mass differences before and after making the methyl ester from the TMS derivative.
C. Method for determining the active hydrogen content of a molecule in a single GC/MS run
This method is especially valuable for identifying metabolites and other trace biological materials. A small portion of the sample is dissolved in the minimum amount of solvent and the solution is divided

into two equal portions. The usual TMS derivative is prepared using the TMS procedure provided previously using BSA reagent (Pierce cat. no. 38838) for one portion. The other portion is used to prepare the deuterated derivative using BSA-D_9 (Merck cat. no. MD-1060). After the reactions are completed, the two portions are mixed and a 1–2 μl sample is injected immediately into the GC/MS system. In general, the front of the GC peak contains the deuterated derivative, while the back of the GC peak contains the regular (D_0) TMS derivative. The mass difference between the MW of the D_0-TMS and the MW of the D_9-TMS derivatives is used to determine the number of TMS groups present. For each 9 Daltons difference, there will be one TMS group. The deuterated TMS derivative will have a fragment ion at m/z 82, while the D_0-TMS derivative will have a fragment at m/z 73. To obtain the MWs, add 18 (CD_3) to the highest mass ion of substantial intensity (excluding isotope peaks) for the D_9-TMS derivative and add 15 (CH_3) for the D_0-TMS derivative. (Note: A molecular ion of low abundance may be present.)

TBDMS Derivative Procedure for Acids

The TBDMS derivative has been used for low molecular weight acids such as formic and for acids such as itaconic, citraconic, and mesaconic where a 30-m DB-210 column programmed from 60 to 210° at 10°/min separates the latter acids as the TBDMS derivatives.

1. Add 50 μl MTBSTFA reagent (Pierce cat. no. 48920) to 1–5 mg dry sample. Solvents that may be used (if necessary) are pyridine, acetonitrile, THF, and DMF.
2. Allow the reaction mixture to react at room temperature for 30 min for low-boiling acids. For high-boiling acids, heat at 60° for 5–20 min.
3. Inject 1– μl into the GC/MS system.

Functional Group

>NH or -NH₂

TMS Derivative	**Increase of MW**
$-NHSi(CH_3)_3$	72

If at least three (CH_2)s are present,

$-(CH_2)_3-N[Si(CH_3)_3]_2$ 144

TBDMS Derivative

$$\underset{\underset{CH_3}{|}}{\overset{\overset{CH_3}{|}}{- NHSiC(CH_3)_3}}$$ 114

Trifluoroacetyl Derivative

$$\overset{\overset{O}{\|}}{- NHCCF_3}$$ 96

Methyl-8 Derivative

$-N=CHN(CH_3)_2$ 55

Acetyl Derivative

$-NHC(O)CH_3$ 42

TMS Procedure

1. Add 50 μl MSTFA (Pierce cat. no. 48911), BSA (Pierce cat. no. 48911), BSA (Pierce cat. no. 38837), or BSTFA (Pierce cat. no. 38832) to the dry sample.
2. Add 50 μl pyridine, acetonitrile, or THF, cap tightly, and mix well.
3. Heat at 60° for 5 min, or let stand at room temperature for 10–20 min.
4. If necessary, cool the reaction mixture to room temperature and inject 1–2 μl into the GC/MS system.

Method to Determine the Number of -NH or -NH$_2$ Groups

1. Run the sample first as the TMS derivative.
2. Add 250 μl MBTFA (Pierce cat. no. 49701) to the TMS reaction mixture.

3. Heat at 60° for 30 min to convert the TMS groups of -NH and -NH$_2$ to trifluoroacetyl groups.
4. Rerun the sample. The deduced MW should increase by 24 Daltons for each -NH or -NH$_2$ present.

TBDMS Procedure

1. Add 50 μl acetonitrile, pyridine, DMF, or other suitable solvent to the dry sample.
2. Add 50 μl MTBSTFA reagent (Pierce cat. no. 48920) plus solvent to the sample and heat at 60° for 15 min.
3. Inject 1–2 μl into the GC/MS system.

Trifluoroacetyl Procedure

1. Add 200–500 μl toluene containing 0.05 M triethylamine to the dry sample.
2. Add 50 μl trifluoroacetic anhydride (TFAA), cap, and mix well.
3. Heat for 5 min at 45°.
4. Cool to room temperature.
5. Add 400–1000 μl 5% sodium bicarbonate solution.
6. Mix by vortexing until the top layer is clear.
7. Centrifuge.
8. Inject 1–2 μl of the top layer. Do not inject any of the bottom layer.

Acetyl Procedure

1. Add 40 μl pyridine and 60 μl acetic anhydride to the dry sample in a septum vial.
2. Heat the reaction mixture at 60° for at least 30 min.
3. Evaporate to dryness with clean, dry nitrogen.
4. Dissolve the residue in 25 μl acetonitrile or pyridine.
5. Inject 1–2 μl into the GC/MS system.

Methyl-8 Derivative (-NH$_2$ groups only)

1. Add 50 μl acetonitrile and 50 μl Methyl-8 reagent (Pierce cat. no. 49350) to <0.1 mg dry sample.
2. Heat at 60° for 30 min.
3. Inject 1–2 μl into the GC/MS.

Functional Group

Methoxime Derivative	Increase of MW per Keto Group

$$\text{>}C = N\text{ - O }CH_3$$

29

Oxime and TMS Formation

$$\text{>}C = N\text{ - O - }Si(CH_3)_3$$

87

Phenylhydrazone Derivative

$$\text{>}C = N \overset{H}{\text{ - N}}\text{ - }C_6H_5$$

102

Methoxime Derivatization Procedure

1. Add 0.5 ml MOX reagent (Pierce cat. no. 45950).
2. Heat at 60° for 3 hr.
3. Evaporate the reaction mixture to dryness with clean, dry nitrogen.
4. Dissolve the residue in the minimum amount of ethyl acetate. Do not add the solvent if the TMS derivative is to be prepared as follows.

TMS Derivative of Methoxime Product

Add 250 μl of MSTFA reagent to the dry methoxime derivative and let stand for 2 hr at room temperature.

Phenylhydrazone Derivative Procedure

(See Knapp, D. R. *Handbook of Analytical Derivatization Reactions.* New York: John Wiley & Sons, 1979.)

Reference

1. McCloskey, J. A., Stillwell, R. N., and Lawson, A. M. "Use of Deuterium Labeled Trimethylsilyl Derivatives in Mass Spectrometry," *Anal. Chem.*, *40*, 233–236, 1968.

A p p e n d i x 4

Cross-Index Chart for GC Phases

Alltech/RSL	Chrompack	HP	J&W	Quadrex	Restek	SGE	Supelco	Maximum temperature
RSL-150 or RSL-160	CP-SIL5 CB	HP-1 or Ultra-1	DB-1	007-1	Rtx-1	BP-1	SPB-1 / SP-2100	300°
RS-L200	CP-SIL 8CB	HP-5 or Ultra-2	DB-5	007-2	Rtx-5	BP-5	SPB-5	300°
OV-1701	CP-SIL 19CB	—	DB-1701	007-1701	Rtx-1701	BP-10	—	260°
RSL-300	CP-SIL 24CB	HP-17	DB-210 / DB-17	007-17	Rtx-50	—	SP-2250	260°
RSL-500	CP-SIL 43CB	HP-225	DB-225	007-225	Rtx-225	BP-225	SP-2300	200°
Superox II	CP-SIL 52CB	HP-20M	DB-Wax	007-CW	Stabilwax	BP-20	Supelcowax 10	220°
Superox FA	FFAP-CB	HP-FFAP	DB-FFAP	007-FFAP	Stabilwax DA	BP-21	NUKOL	220°
RSL-1000	CP-SIL-88	—	DB-23	—	Rtx-2330 / Rtx-2340	BPX-70	SP-2330 / SP-2340	230°
AT-624	CP13CB for halocarbons	—	DB-624	007-624	Rtx for volatiles	—	VOCOL	260°
Pesticide column	CP-SIL 8CB for pesticides	—	DB-608	007-608	Rtx-35	—	SPB-608	300°
CP-cyclodextrin-2,3,6-M-19	—	—	Cyclodex B	—	—	—	—	200°

Appendix 5

McReynolds' Constants

GC phase	X'	Y'	Z'	U'	S'
DB-1	16	55	44	65	42
DB-5	19	74	64	93	62
DB-1701	67	170	153	228	171
DB-210	146	238	358	468	310
DB-17	119	158	162	243	202
DB-225	228	369	338	492	386
DB-Wax	322	536	368	572	510
DB-FFAP	340	580	397	602	627
DB-23	386	610	506	710	591

Source: McReynolds, W. O. *J. Chromat. Sci.*, *8*, 685, 1970.

Appendix 6

Simple GC Troubleshooting

In GC troubleshooting,[1-3] it is always good practice to consult the manufacturer's guide for the instrument. If this is unavailable, incomplete, or does not help solve the problem, you may want to read further before calling for service. Also, it is wise to run test samples regularly on a test column such as a 30-m DB-1 or DB-5. By doing test samples, it is possible to check the performance of the GC by running the test mixture. A good test mixture is n-$C_{14}H_{30}$, n-$C_{15}H_{32}$, and n-$C_{16}H_{34}$, which can be obtained from Supelco. Always run the test sample under the same conditions such as 100–275° at 10°/min and with the same linear gas velocity. Sharp well-defined peaks usually mean the system is operating properly. If the GC peaks are tailing or otherwise distorted and/or the retention time changes, then there is a problem that must be corrected.

Problem	Possible Cause
Tailing peaks	• The column is deteriorating (often caused by water or oxygen in the carrier gas). • Injection port and/or column temperature is too low. • Sample components adsorb in the injection port, in the transfer line, or on the column.

Problem	Possible Cause
Poorly resolved peaks that were previously resolved	• Linear velocity of carrier gas is wrong. • Columns or temperatures are wrong. • Column is contaminated or deteriorated.
Retention times longer or shorter on the same column	• Column temperature is too high or too low. • Carrier gas flow is incorrect. • The septum and/or column leaks.
GC peak broadening	• Thick film column is used. Thick film should only be used with volatile or reactive samples. • Injection port temperature is too low.
Extra peaks not present in the sample chromatogram	• Component peaks from previous run that are usually too broad for that retention time are present. Rerun the sample and run the analysis longer. • Impurities from solvents, reagents, or sample vials are present.

When a problem is noted, it is generally best to look for solutions in easy-to-check-out areas. For example, check for leaks at all column connections and external fittings. Once the simple checks are completed, attempt to isolate the problem to the injector, column, detector (or transfer line), or instrument.

The *column* can be checked for problems by replacing it with one of known performance. If the problem is eliminated, reinstall the original column, checking for proper insertion distances to see if the problem reappears. If it does not, the problem was likely corrected during reinstallation.

Injector problems are often associated with contaminated inlet sleeves, a leaking or dirty septum, or a leaking inlet O-ring or ferrule seal. Replace spent gas purifiers, and check flow rates, linear velocity, and inlet temperature, and for incorrect sleeve type (e.g., splitless sleeve for split analysis), a bad inlet solenoid valve, and so on. The injector is implicated if the column can be switched to a second detector and the problem persists. Replacing the column with a few meters of deactivated fused silica tubing connected to the FID detector with the linear gas velocity adjusted to 20–40 cm/sec can be used to detect injector problems. The injector is implicated if, on injecting the same solvent used with the problem sample and running the temperature program, a noisy baseline is observed or peaks are detected.

Because two *detectors* are connected simultaneously in GC/MS, it is possible to rule out a detector problem if the erratic result occurs on both detectors (e.g., MS, FID). A poor result at one detector could be associated with a postcolumn splitter or a leak in the transfer line. Once the problem is isolated, the instrument manual is usually a valuable source for information on possible fixes.

Contamination Problems

Frequently, column problems are caused by the samples that are being analyzed. This type of problem is more likely to occur on capillary columns because of their low capacity for contamination. Contamination results when the sample contains nonvolatile or even semivolatile materials such as salts, sugars, proteins, and so on. Column contamination is more frequently observed with splitless injection because larger amounts of material are being injected on the column.

The symptoms of column contamination include irregular peak shape, loss of resolution, loss of retention, irregular or noisy baseline, and ghost peaks from semivolatile materials of a previous run or from sample decomposition. Some of these problems can be the result of a contaminated injector.

One solution is to replace the column, but a less expensive approach is to attempt to clean the column. Baking is one approach that removes some forms of contamination, but also shortens the column life because it removes some of the stationary phase. A solvent rinse is the most effective means of cleaning a bonded or cross-linked phase column. Solvent rinse kits are available with instructions from most column manufacturers. The procedure involves forcing solvents through the GC column, usually in the following order—water, methanol, methylene chloride, and hexane—using 10–15 psi back pressure.

The injector should also be cleaned by removing the liner and cleaning the cooled metal injector with methanol. Allow the injector to dry and insert a clean liner.

Troubleshooting the GC/MS Interface

The best practice in troubleshooting an interface is first to determine that the problem is in the interface. If upon connecting the GC column to an alternate detector, the problem is no longer evident, then it is likely an interface problem. Problems with capillary columns usually involve column plugging. This problem can be alleviated by breaking off a small section at the front of the column. Because plugging can be caused by a cold spot in the transfer line, make certain that the interface heaters are working properly.

Packed and macrobore columns require enrichment devices and, therefore, have more possibilities for failure. If the interface heaters are working properly, then the most likely problem is a leak at one of the connections. A helium detector can be used to hunt for leaks, but unfortunately in most interface designs this requires cooling the line before leak detecting. A better method is to direct a small flow of argon at interface connections while monitoring for *m/z* 40 on the mass spectrometer. Another problem source is fouling of the jet separator. Replacing the jet separator will solve this problem.

References

1. Rood, D. *A Guide to the Care, Maintenance and Troubleshooting of Capillary Gas Chromatographic Systems*. Heidelberg, Germany: Huethig Publishing, 1991.
2. *Supelco Trouble Shooting Guide 792*. Supelco, Inc., Supelco Park, Bellefonte, PA 16823-0048
3. Restek Catalog # 20450. Restek Corporation, 110 Benner Circle, Bellefonte, PA 16823-8812.

Appendix 7

Definitions of Terms Related to Mass Spectrometry*

Atmospheric pressure chemical ionization (APCI): Chemical ionization at atmospheric pressure.

Base peak: The peak in the mass spectrum with the greatest intensity.

Collision-activated dissociation (CAD): The same process as collision-induced dissociation (CID).

Collision-induced dissociation (CID): An ionic/neutral process in which the projectile ion is dissociated as a result of interaction with a target neutral species. Part of the translational energy of the ion is converted to internal energy causing subsequent fragmentation.

Chemical ionization (CI): The formation of new ionized species when gaseous molecules interact with ions. This process may involve the transfer of an electron, proton, or other charged species between the reactants in an ion–molecule reaction. CI refers to positive ions, and negative CI is used for negative ions.

Ionization: A process that produces an ion from a neutral atom or molecule.

Detection limit: The detection limit is the smallest sample flow that provides a signal that can be distinguished from background noise.

Electron impact ionization (EI): Ionization by electrons (M + e$^-$ → M$^+$ + 2e$^-$).

Electrostatic analyzer (ESA): A velocity-focusing device for producing an electrostatic field perpendicular to the direction of ion travel. Ions of a

*For a complete list of definitions, see: Price, P. *J. Am. Soc. Mass Spectrom.*, 2, 336, 1991.

given kinetic energy are brought to a common focus. This analyzer is used in combination with a magnetic analyzer to increase resolution and mass accuracy.

GC/MS interface: An interface between a GC and a MS that provides a continuous introduction of effluent gas to the MS ion source.

Ion cyclotron resonance (ICR) analyzer: A device to determine the mass-to-charge of an ion in a magnetic field by measuring its cyclotron frequency.

Ion trap analyzer: A mass-resonance analyzer that produces a three-dimensional rotationally symmetric quadrupole field capable of storing ions at selected mass-to-charge ratios.

Isotopic ion: An ion containing one or more of the less abundant naturally occurring isotopes of the elements that make up the structure.

Magnetic analyzer: A direction-focusing device that produces a magnetic field perpendicular to the direction of ion travel. All ions of a given momentum with the same mass-to-charge ratio are brought to a common focus.

Mass spectrometer (MS): An instrument used to analyze ions according to their mass-to-charge ratios to determine the abundance of ions.

Mass-to-charge ratio (m/z): An abbreviation for division of the observed mass of an ion by the number of charges the ion carries. Thus, $C_6H_6^+$ has m/z 78, but $C_6H_6^{+2}$ has m/z 39.

Metastable ion: An ion that dissociates into a product ion and a neutral product during the flight from the ion source to the detector. Metastable ions can be observed when they occur in a field-free region of the MS.

Molecular ion: An ion formed by addition (M^-) or removal (M^+) of an electron from a molecule without fragmentation.

Monoisotopic mass: The mass of an ion calculated using the exact mass of the most abundant isotope of each element in the formula (e.g., C = 12.0000, O = 15.9949).

Nominal mass: The mass calculated for an ion when using the integer mass values of the most abundant isotope of each element in the formula (e.g., C = 12, O = 16, S = 32).

Parent ion/precursor ion: An ion that undergoes fragmentation to produce a daughter ion and a neutral product.

Product ion/daughter ion: An ion related to a precursor or parent ion by a process such as fragmentation. For example, a parent ion fragments to produce a daughter ion.

Quadrupole analyzer: A mass filter that creates a quadrupole field with dc and rf components so that only ions of a selected mass-to-charge are transmitted to the detector.

Radical ion (odd electron ion): An ion containing an unpaired electron that is both a radical and an ion.

Rearrangement ion: A dissociation product ion in which atoms or groups of atoms have transferred positions in the parent ion during the fragmentation process.

Resolution (10% valley definition): Let two peaks of equal height in a mass spectrum at masses m and (m − Δm) be separated by a valley that at its lowest point is 10% of the height of the peaks.

Selected ion monitoring (SIM): Describes the operation of a MS in which the ion currents of one or several selected m/z values are recorded, rather than the entire mass spectrum.

Total ion current (TIC): The sum of all the separate ion currents contributed by the ions that make up the spectrum.

Time-of-flight (TOF) analyzer: A device that measures the flight time of ions having a given kinetic energy through a fixed distance. The flight times of ions are proportional to their mass-to-charge ratio.

Appendix 8

Tips and Troubleshooting
for Mass Spectrometers

Logbook Maintenance

A logbook should be maintained on the operation of each MS. In this book, the instrument tuning parameters used to obtain reference mass spectra should be kept. The tuning parameters should include the repeller, lens, and multiplier voltages or settings as well as other parameters, such as the gain setting, instrument resolution, source temperature, scan speed, accelerating voltage, and so forth. If the instrument is autotuned, print out the autotune report. It is best to obtain the spectra when the instrument has just met manufacturer specifications and after servicing. The lens settings for optimum performance will change with time and the multiplier gain may need to be increased, but this data set makes it possible to follow the performance of the instrument and readily determine when the cause of unsatisfactory GC/MS data is due to the MS. Deteriorating performance is most often an indication that the ion source is dirty.

Simple Troubleshooting

The logbook should also contain a description of problems that are encountered with the MS and the fix that was employed. Over time the same symptoms are likely to recur. Occasionally, add a reference spectrum to the logbook with the parameters used to obtain the spectrum. This will give an indication of the average operating performance of the instrument.

A slow deterioration that is not remedied by cleaning the ion source and not related to the multiplier gain is an indication that the analyzer needs to be cleaned. For quadrupole instruments, cleaning the prefilter is often sufficient to restore performance. (A prefilter is not available on all quadrupole instruments.)

The most likely cause of problems in any MS is associated with the ion source. When a MS problem is discovered, a good rule is to clean the source and check for shorts or a burned-out filament. For most instruments, it is relatively simple to determine that a filament has burned out because the emission current will be zero. Sometimes, the filament will short to the block. In this case, the emission current will read high when in the CI mode where emission current is read between the filament and the ion source block. In either case, the source should be cleaned and the filament inspected. If the filament is sagging, it is a good idea to replace it.

Air leaks are another source of trouble in the MS. A simple method of leak detection is to squirt a small volume of acetone on flanges and other areas where leaks could occur. *Caution is advised not to use this procedure near hot surfaces because of the flammability of acetone.* A second way to test for small leaks is to tune the MS to m/z 40 and to use argon to test for leaks. The m/z 40 peak will increase if argon enters the source. Helium (m/z 4) is a better choice, except when helium carrier gas is used in conjunction with the GC. A small stream of the gas is aimed at all seals where a leak can occur. If a leak is detected at a seal, it can sometimes be stopped by tightening the seal, but it is better to replace the seal than to overtighten it.

Calibration

Calibration is another source of potential problems. The frequency with which an instrument needs to be calibrated for mass accuracy depends on the instrument. For some instruments, it might be necessary to recalibrate every day and check the calibration at least once a day. Other instruments will hold calibration for months. A good suggestion and a good laboratory practice for all instruments is to run a quick calibration check each morning. This can be as simple as injecting a fixed quantity of a volatile solvent into the GC. A quick glance at the resulting mass spectra will confirm that the mass assignment is correct. This procedure also allows some assessment of the total performance of the GC/MS instrument and only takes a few minutes.

In the positive ion mode, calibration is most frequently performed using perfluorokerosine, but different instrument manufacturers recommend different reference materials. Consult your instrument manual for the recommended procedure. The calibration reference masses will be stored in a calibration file.

Accurate Mass GC/MS

Identification of unknowns using GC/MS is greatly simplified if accurate mass measurements are made of all the ions in a spectrum so that reasonable elemental compositions of each ion are available. Unfortunately, obtaining a mass measurement that is accurate enough to significantly limit the number of possible elemental compositions requires expensive instrumentation such as a double-focusing magnetic sector or fourier transform ICR MS.

Accurate mass measurement is generally associated with high resolution. Mass resolution is necessary to eliminate interference of either the ions whose masses will be measured or the internal reference ions by other ions appearing at the same nominal mass. If good chromatography techniques are achieved, interferences from ions of compounds other than the compound being mass measured are essentially eliminated and the chief cause of interference problems is the overlap of ions from the sample and internal reference. For this reason, an internal reference material is chosen in which all the peaks have as large a negative mass defect as possible. If perfluoroalkanes are used, a resolution of more than 3000 ($M/\Delta M$, 10% valley definition) is usually necessary to eliminate peak overlap and poor results.

Because an increase in resolution causes a decrease in sensitivity, it is best to operate at the lowest resolution commensurate with good results. Some instrument data systems will allow calibration with an external reference material such as perfluorokerosene and then use of a secondary reference material for the internal mass reference. Tetraiodothiophene, vaporized using the solids probe inlet, is recommended as the secondary reference. The accurate masses are 79.9721, 127.9045, 162.9045, 206.8765, 253.8090, 293.7950, 333.7810, 460.6855, and 587.5900. For a higher mass standard, use hexaiodobenzene. Because the mass defect for these internal reference ions are so large, a resolution of 2000 is ample to separate these ions from almost any sample ions encountered in GC/MS.

Negative Chemical Ionization GC/MS

Negative CI can give excellent results for certain types of compounds. Compounds with electronegative substituents and unsaturation can be expected to have a large electron capture cross-section and thus work well in the negative ion mode. Frequently, much higher sensitivity is obtained for these compound types in the negative ion mode than under positive ion conditions. In addition, the molecular ion is usually very abundant. The method requires that a gas such as nitrogen or argon be added to the source at a slightly lower pressure than is used in positive ion CI. It is best to tune the ion source pressure and source parameters using a peak from the reference material.

Calibration in negative ion mode to at least m/z 700 can be achieved using perfluorokerosine. Polyperfluoroisopropylene oxide oligomer (see following structure) distributions have been used to calibrate to m/z 5000 in the negative ion electron attachment mode. To reach m/z 5000,* it was necessary to vaporize the oligomer from the solids probe. This material can be used as an internal reference for accurate mass negative ion GC/MS. The reference ions are: 31.9898, 68.9952, 84.9901, 116.9963, 162.9818, 184.9837, 250.9754, 282.9817, 328.9671, 350.9691, 416.9607, 448.9670, 494.9524, 516.9544, 614.9524, 682.9397, 780.9377, 848.9251, 946.9230, 1014.9104, 1112.9084, 1180.8957, 1278.8937, 1346.8811, 1444.8790, 1512.8664, 1610.8643, 1678.8517, 1776.8496, 1844.8370, and 1942.8349.

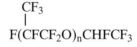

$$\underset{\displaystyle \mathrm{F(CFCF_2O)_nCHFCF_3}}{\overset{\displaystyle \mathrm{CF_3}}{\overset{\displaystyle |}{}}}$$

*The material used for this work was Freon E-13. The 13 refers to the most abundant oligomer ($n = 13$) in the above structure.

A p p e n d i x 9

Atomic Masses and Isotope Abundances

Nuclide	Mass	Abundance (%)
^{1}H	1.0078	100
^{2}H	2.0141	0.015
^{10}B	10.0129	20
^{11}B	11.0093	80
^{12}C	12.0000	98.9
^{13}C	13.0034	1.1
^{14}N	14.0031	99.6
^{16}O	15.9949	99.8
^{19}F	18.9984	100
^{28}Si	27.9769	92.2
^{29}Si	28.9765	4.7
^{30}Si	29.9738	3.1
^{31}P	30.9738	100
^{32}S	31.9721	95
^{34}S	33.9679	4.2
^{35}Cl	34.9689	75.8
^{37}Cl	36.9659	24.2
^{79}Br	78.9184	50.5
^{81}Br	80.9164	49.5
^{127}I	126.9047	100

Appendix 10

Structurally Significant McLafferty Rearrangement Ions

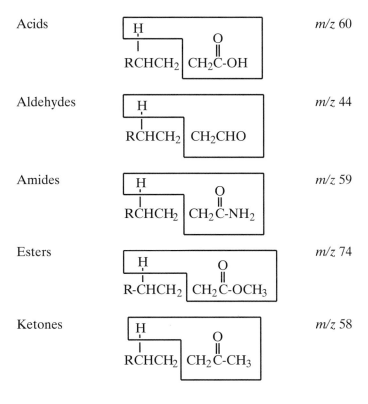

Acids m/z 60

Aldehydes m/z 44

Amides m/z 59

Esters m/z 74

Ketones m/z 58

Nitriles

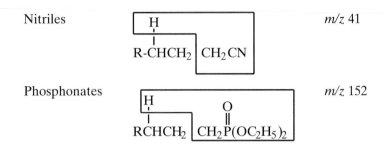

m/z 41

Phosphonates

m/z 152

Appendix 11

Isotope Patterns for Chlorine and Bromine

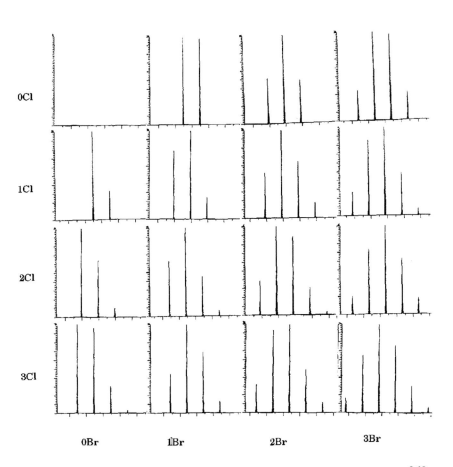

0Cl

1Cl

2Cl

3Cl

0Br   1Br   2Br   3Br

Appendix 12

Mixtures for Determining Mass Spectral Resolution

Mixtures for determining resolution	Ratio		Resolution (M/ΔM, 10% valley)
Bromobenzene/dimethylnaphthalene	1:1	*m/z* 156	1144
Cyclohexane/cyclopentanone	2:1	*m/z* 84	2300
Styrene/benzonitrile	3:1	*m/z* 103	8195
200° inlet temperature required			
Dimethylquinoline/dimethylnaphthalene	1:2	*m/z* 156	12,400
1-Octene	—	*m/z* 71	15,900
Toluene/xylene	1:10	*m/z* 92	20,922

Part IV References

Gas Chromatography Manufacturers and Suppliers

"Gas Chromatography Manufacturers or Suppliers," *J. Chromatogr. Sci.*, *32*, 10G, 1993.

Analytical Columns and Supplies Catalog. HP Analytical Direct, 2850 Centerville Road, Wilmington, DE 19808.

Chiraldex Capillary GC Columns "Chiral Separations." ASTEC, 37 Leslie Court, P.O. Box 297, Whippany, NJ 07981.

Chromatographic Research Supplies Catalog. Chromatographic Research Supplies, P.O. Box 888, Addison, IL 60101.

Pierce Handbook and General Catalog "Derivatizing Agents for GC/MS." Pierce, P.O. Box 117, Rockford, IL 61105.

The Reporter. Supelco, Supelco Park, Bellefonte, PA 16823. (For example, see Vol. 13, No. 2, 1994.)

The Restek Advantage and *Chromatography Products Catalog*. Restek Corporation, 110 Benner Circle, Bellefonte, PA 16823-8812. (For example, see Hints for Capillary Chromatography, Vol. 3, No. 3, p. 12, 1992.)

Accurate Mass Measurement GC/MS

Grange, A. H., Donnelly, J. R., Brumley, W. C., Billets, S., and Sovocool, G. W. *Anal. Chem.*, *66*, 4416, 1994.

Location of Double Bonds in Aliphatic Compounds

Murphy, R. C. *Mass Spectrometry of Lipids*. New York: Plenum Press, 1993. An extensive review of techniques for locating double bonds in unsaturated fatty acids.

Nakamura, T., Takazawa, T., Maruyama-Ohki, Y., Nagaki, H., and Kinoshita, T. *Anal. Chem.*, *65*, 837, 1993.

Schneider, B. and Budzikiewicz, H. *Rapid Commun. Mass Spectrom.*, *4*, 550, 1990.

Novel Derivatives for GC/MS

Kudzin, Z. H., Sochacki, M., and Drabowicz, J. *J. Chromatogr.*, *678*, 299, 1994.

Stable Isotope Dilution GC/MS

Allen, M. S., Lacey, M. J., and Boyd, S. *J. Agr. Food Chem.*, *42*, 1734, 1994.

Index